Cooperative Extension

Enhancing Wildlife Habitats
A Practical Guide for Forest Landowners

Authors

Scott S. Hobson
Chief Biologist
Keyes Associates and Environmental Scientific
Lincoln, Rhode Island

John S. Barclay
Associate Professor of Wildlife Management
Extension Wildlife Specialist
University of Connecticut

Stephen H. Broderick
Extension Forester
University of Connecticut

NRAES—Natural Resource, Agriculture, and Engineering Service
Cooperative Extension
PO Box 4557
Ithaca, New York 14852-4557

NRAES–64
October 1993

© 1993 by the Natural Resource, Agriculture, and Engineering Service
All rights reserved. Inquiries invited. (607) 255-7654

ISBN 0-935817-35-2

About NRAES

NRAES, the Natural Resource, Agriculture, and Engineering Service, is a not-for-profit program dedicated to assisting land grant university faculty and others in increasing the public availability of research- and experience-based knowledge. NRAES is sponsored by eleven land grant universities in the eastern United States. We receive administrative support from Cornell University, the host university.

When you buy books from NRAES, you are helping to improve the accessibility of land grant university knowledge. While 15% of NRAES' annual income is provided by member universities, the funds to publish new books and coordinate new conferences come from our customers through book sales, conference registrations, and occasional project-specific grants.

NRAES publishes practical books of interest to fruit and vegetable growers, landscapers, dairy and livestock producers, natural resource managers, SWCD (soil and water conservation district) staff, consumers, landowners, and professionals interested in agricultural waste management and composting. NRAES books are used in cooperative extension programs, in college courses, as management guides, and for self-directed learning.

NRAES publishes two types of books: peer-reviewed books and conference proceedings. Our peer-reviewed books are evaluated prior to publication for technical accuracy and usefulness to the intended audience. The reviewers may include university faculty, extension educators, potential users, and interested persons from government and agribusiness. Conference proceedings are not peer-reviewed. However, the authors of papers presented at NRAES-sponsored conferences are chosen for their recognized expertise.

Requests to reprint parts of this publication should be sent to NRAES. In your request, please state which parts of the publication you would like to reprint and describe how you intend to use the reprinted material.

Contact NRAES for more information or a free book catalog.
Natural Resource, Agriculture, and Engineering Service (NRAES)
Cooperative Extension, PO Box 4557
Ithaca, New York 14852-4557
Phone: (607) 255-7654 • Fax: (607) 254-8770 • E-mail: nraes@cornell.edu • Web site: www.nraes.org
Marty Sailus, NRAES Director

NRAES is sponsored by these Land Grant Universities:
University of Connecticut
University of Delaware
University of the District of Columbia
University of Maine
University of Maryland
University of New Hampshire
Rutgers University
Cornell University
University of Vermont
Virginia Polytechnic Institute & State University
West Virginia University

Table of Contents

Figures and Tables
List of Figures vi
List of Tables x

Preface
Preface xi

1 Basic Forest Wildlife Ecology

Introduction 1
The Growth of Forests 2
    Forest Succession 2
    Special Features of Forest Succession 3
The Forest Community 5
    The Food Pyramid 6
    Food Chains and Food Webs 7
    The Role of Decomposers 8
    Niche 9
Wildlife Populations 10
    Carrying Capacity 11
    Reproduction 11
    Mortality 12
Review Questions 15
Field Exercises 16

2 Understanding Wildlife Habitats

Introduction 17
Habitat Requirements of Wildlife 17
    Cover 18
    Special Examples of Cover 19
    Food 21
    Special Examples of Food Sources 22
    Water 23
    Special Examples of Water Sites 24
Additional Habitat Components to Consider 25
    Edge and Ecotone 25
Wildlife Movements 27
    Home Range 27
    Territoriality 27
    Dispersal 28
    Migration 28
    Travel Lanes 28
    Special Example of Travel Lanes 29

Table of Contents

Observing Wildlife and Recognizing Signs of Presence 29
    Tracks 31
    Feeding Evidence 32
    Scats 32
    Other Signs of Wildlife Presence 32
Review Questions 35
Field Exercises 36

3 American Woodcock and Ruffed Grouse

Introduction 38
American Woodcock 39
    Description and Range 39
    Life History 40
    Habitat Requirements 44
    Managing Habitats for Woodcock 47
Ruffed Grouse 53
    Description and Range 53
    Life History 54
    Habitat Requirements 59
    Managing Habitats for Ruffed Grouse 62
Review Questions 67
Field Exercises 68

4 White-tailed Deer and Eastern Wild Turkey

Introduction 69
White-tailed Deer 70
    Description and Range 70
    Life History 71
    Habitat Requirements 77
    Managing Habitats for White-tailed Deer 79
    Deer as a Nuisance 83
Eastern Wild Turkey 86
    Description and Range 86
    Life History 87
    Habitat Requirements 91
    Managing Habitats for Eastern Wild Turkeys 93
Review Questions 98
Field Exercises 99

5 Other Upland Forest Wildlife Species

Introduction 100
Selected Species Groups 101
    Rabbits and Hares 101
    Tree Squirrels and Chipmunks 102
    Woodchucks 103
    Small Mammals 103
    Canids 105
    Felids 106
    "Upland" Mustelids 107

    Songbirds and Woodpeckers *108*
    Raptors *109*
Managing Habitats for Upland Forest Wildlife *110*
    Maintaining Early Successional Habitats for Food and Cover *111*
    Managing Pole- and Timber-sized Forests for Food and Cover *114*
    Trail Systems *120*
Review Questions *124*
Field Exercises *125*

6 Wetlands Wildlife

Introduction *126*
What Are Wetlands? *127*
    Definition of Wetlands *127*
    Description of Wetland Types *127*
    Inland Wetlands *128*
    Significance of Inland Wetlands to Wildlife *133*
    The Role of Inland Wetlands *133*
Selected Species *133*
    Wood Duck *133*
    Other Cavity-nesting Waterfowl *135*
    American Black Duck *135*
    Other Ground-nesting Waterfowl *136*
    Herons and Bitterns *137*
    Rails and Grebes *138*
    Muskrat *139*
    Beaver *141*
    Aquatic Mustelids *142*
    Raccoon *142*
Managing Forest Wetland Habitats for Wildlife *142*
    Maintaining Buffer Strips and Corridors *143*
    Measures to Prevent Erosion and Sedimentation *144*
    Managing Cover and Food Plants *144*
    Managing for Cavity Trees, Snags, and Beaver *146*
    Artificial Nest Structures for Inland Wetlands *147*
    Trail Systems *148*
    A Note on Pond Design *148*
Review Questions *150*
Field Exercises *151*

Appendix

Table 9. List of scientific names of plants mentioned in the text. *152*
Table 10. List of scientific names of animals mentioned in the text. *156*
Table 11. Metric conversions. *159*

Glossary

Glossary *160*

References

Literature Cited & Selected References *165*

List of Figures

1 Basic Forest Wildlife Ecology

1. Only shade-tolerant trees and shrubs can thrive in the understory of a dense forest. 2
2. Each stage of forest succession is important to some wildlife species. 4
3. Birds, mammals, water, and wind all carry forest seeds to distant places. 5
4. Forest types and the wildlife that live in them vary across the northeast region of the United States. 6
5. The number of species declines rapidly with each step up the food pyramid. 7
6. Oak leaves, gypsy moth caterpillars, northern orioles,
   and sharp-shinned hawks are all links on the same food chain. 8
7. Every species has its special role, or niche, in the forest environment. 9
8. Some species of wildlife have developed niches which are specialized and unique. 10

2 Understanding Wildlife Habitats

9. All wildlife need some form of protective cover. 18
10. Rock outcrops are an additional source of shelter for many animals. 19
11. Tree cavities provide protective cover for many wildlife species. 20
12. The woodpecker creates cavities as it feeds. 20
13. Dead, standing trees called snags contribute food, cover, and diversity to the forest environment. 21
14. Maintaining a variety of native mast-producing trees and shrubs
    is an important part of wildlife management. 22
15. Wolf trees, which develop in the open without competition, are valuable for food and cover. 24
16. Ecotones, i.e. transition zones between different plant communities,
    can be especially productive wildlife habitat. 24
17. This aerial view of a variety of plant communities and land uses
    illustrates interspersion and edge habitats. 25
18. Possible configurations of forest vs. disturbed area ecotones on the same parcel. 26
19. Some animals, such as deer, often rely on familiar pathways (trails) from one area to another. 29
20. a. Hemlock cone fragments on the snow are a feeding sign of the red squirrel. 30
    b. Red maple sprouts are browsed by white-tailed deer. 30
    c. Gray squirrels dig in the snow for food. 30
    d. The white-tailed deer leaves clumps of droppings (pellets). 30
    e. Otter tracks typically appear paired as a result of the loping or bounding gait of the animal. 31
    f. Coyote tracks are more pointed than those of the dog or fox. 31
    g. Pheasant and deer have passed through the same area. 31
    h. Male turkey tracks show a long middle toe. 31

3 American Woodcock and Ruffed Grouse

21. The woodcock is the only member of the sandpiper family adapted to forest
    rather than shoreline habitats. 39
22. Woodcock have been known to breed as far north as central Canada
    and to winter as far south as central Florida. 40
23. Woodcock are best known for their unique spring courtship displays. 41
24. The woodcock's nest is simply a shallow depression in the leaf litter,
    located close to a sapling or other guard object. 42

25	The woodcock's bill is adapted for probing the soil and grasping underground food.	43
26	Singing grounds are a critical habitat component. Clearings in brushy fields close to moist soils make the best sites.	44
27	Preferred nesting sites are in young, open woodlands with a scattered understory of brush and seedlings.	45
28	Dense alder stands make excellent woodcock feeding sites.	46
29	Probe holes are clear evidence that woodcock have been feeding in an area.	47
30	Woodcock are most attracted to singing grounds of at least 500 square feet, managed to limit the height of trees and shrubs around the perimeter.	48
31	Feeding sites are best managed on twenty- to twenty-five-year rotations, to maintain brushy patches of varying ages.	50
32	As feeding sites approach maturity, their suitability for nesting and broad rearing increases.	51
33	Pastures and lowbush blueberry fields make excellent roosting areas.	51
34	Ruffed grouse grow a temporary fringe on their toes for "snowshoeing" during the winter.	52
35	Ruffed grouse are widely distributed throughout much of Canada and the northern United States.	53
36	The "drumming" of the male ruffed grouse creates a distinctive and memorable sound.	53
37	The raised ruff of the courting male helped give the ruffed grouse its name.	54
38	Chicks leave their nest shortly after hatching and begin to fly within a week.	55
39	Grapes, rose hips, and other "soft mast" are important seasonal grouse foods.	56
40	For reasons not fully understood, ruffed grouse populations tend to fluctuate in cycles of about ten years.	57
41	Central tail feathers of ruffed grouse may be used in determining sex.	58
42	Drumming sites are typically located within brushy open woodlands in intermediate stages of forest succession.	59
43	Clearings or sparse woodlands with an abundant herbaceous ground cover are preferred sites for brood rearing.	60
44	Droppings on a large downed log may indicate an established grouse drumming site.	62
45	Tracks and snow roosts are telltale winter signs of grouse.	63

4 White-tailed Deer and Eastern Wild Turkey

46	The white underside of the tail communicates alarm when the deer is threatened.	70
47	The Northern Woodland Whitetail is the most widely distributed subspecies of deer in the United States.	71
48	Fawns spend little time with the doe for the first two weeks, remaining concealed among ground cover while the doe feeds elsewhere.	73
49	Winter feeding is often limited to browse and whatever mast crops remain available during snow cover.	74
50	Heavy feeding can result in a "browse line" approximately 6 feet high, below which all palatable forage has disappeared.	75
51	The transition zone between field and forest can be an excellent feeding habitat.	76
52	Groups of pellet-shaped droppings like the one pictured are sure signs of deer presence.	78
53	Deer may use favored routes repeatedly until trails become evident.	79
54	The shape of droppings can provide clues to what deer are feeding on.	79
55	Bucks test their antlers and release aggression by rubbing against saplings.	80
56	Removing large trees adjacent to woods, roads, and trails can stimulate the development of woody browse and herbaceous growth.	81
57	Undesirable trees such as maturing hardwoods in yarding areas can be killed by girdling. Such trees form snags which are beneficial to many wildlife species.	82
58	Ideal winter cover includes corridors of conifers which extend into areas of abundant browse.	84
59	The range of wild turkey is expanding due to changing land use and restoration efforts.	86
60	The displaying male gobbles and struts with tail fanned, feathers puffed, head back, and wings lowered to the ground.	88

List of Figures

61 The male turkey is larger than the female, with a beard protruding from the breast; red, fleshy caruncles and dewlap on the head and neck; and a snood which hangs from above the bill. *90*
62 The foot of the adult male wild turkey differs from that of the female in that it is larger, has a middle toe that is longer than the other two lateral toes, and has a sharp spur above the foot. *92*
63 Wild turkey poults, such as these 3½-month-old youngsters, feed avidly on insects found on herbaceous plants in forest openings and along field borders. *94*

5 Other Upland Forest Wildlife Species

64 Available woody browse just above winter snow lines is crucial to the survival of rabbits and hares. *101*
65 Tree squirrels are opportunistic feeders known to consume eggs, insects, and even nestling birds along with hard and soft mast. *102*
66 Bats require well-sheltered areas for cover, such as hollow trees, ledges, or caves. *104*
67 The fox and coyote are predatory mammals. A good habitat for small mammals is a good habitat for these canids. *105*
68 Bobcats are secretive predators that occupy the highest trophic levels in the food chain. *106*
69 Weasels prefer early successional forests with an abundance of clearings, fields, and brushy edges. *107*
70 Each songbird species occupies a specific niche which separates its feeding habits from those of other species. *109*
71 Trees with large cavities are important nesting sites for owls. *110*
72 Herbaceous openings contain abundant insect populations and are important seasonal habitat for many species. *111*
73 Openings containing young woody brush provide additional, equally valuable food and cover. *111*
74 Five- to ten-year cutting rotations maintain brushy openings in various stages of development, increasing their habitat value. *112*
75 Wide ecotones like the one in the background can be created by leaving an unmowed strip around a field's edge. *114*
76 Forests with high wildlife species diversity generally contain canopies of different heights within the same area. *115*
77 These 1–2 acre patch cuts create early successional habitat in an otherwise uniformly mature forest. *116*
78 Trees that are uprooted and/or blown over provide added cover which is particularly valuable in woodlands lacking a dense understory. *119*
79 Where trails follow steep or lengthy slopes, diversions must be built to channel running water off the trail. *120*

6 Wetlands Wildlife

80 Example of a stunted red maple swamp. *127*
81 Forested wetlands with sparse overstory and lush undergrowth. *128*
82 Vernal pool in woodland with light snowcover. *129*
83 Shallow marsh with dense vegetation bordering deep marsh. *130*
84 Blue flag, a common resident of wet meadows. *131*
85 Northern New England spruce bog. *131*
86 Deepwater habitat bordered by deep and shallow marshes. *132*
87 Depiction of the Atlantic flyway. *134*
88 Breeding and wintering ranges of the black duck in eastern North America. *136*
89 Two young great blue herons stand by their nest in the rookery. Nearby an adult great blue heron sits on a nest in the crotch of trees. *138*
90 Some New England marsh birds: king rail, Virginia rail, pied-billed grebe, and sora. *139*
91 Muskrat lodge in marsh. Note herbaceous plant materials used. *140*
92 Beaver lodges appear as mounds of logs, branches, and mud. *141*

93 Careful planning of haul road and/or skid trail layout will minimize stream and wetland impacts from logging. *143*
94 Transitional vegetation from open water to marsh and forest. *145*
95 Just hatched downy duckling in wood duck nestbox opening. *146*
96 Nesting platform for Canada goose, showing construction detail. *147*

List of Tables

1 Basic Forest Wildlife Ecology

1. Characteristics of common northeastern trees and woody shrubs. *13*
2. Relative food values of selected northeastern trees and shrubs for wildlife and for honey bees by stage of forest development. *14*

2 Understanding Wildlife Habitats

3. Some seed characteristics of commercially important North American forest trees. *33*
4. Status, relative abundance, and home range requirements of selected New England mammalian species. *34*

5 Other Upland Forest Wildlife Species

5. Summary of forest habitat characteristics required by selected forest interior breeding birds. *121*
6. Raptors that breed in New England. *122*
7. Number of cavity trees needed to sustain the hypothetical maximum populations of nine species of woodpeckers found in New England. *123*

6 Wetlands Wildlife

8. An example of recommended buffer strip widths (in feet). *149*

Appendix

9. List of scientific names of plants mentioned in the text. *152*
10. List of scientific names of animals mentioned in the text. *156*
11. Metric conversions. *159*

Preface

Some 70% of all the woodland in the northeastern U.S. belongs to private, non-industrial landowners. It follows, therefore, that the majority of the region's terrestrial wildlife depends upon these private owners to provide quality habitat: the combination of food, water, and shelter they need to reproduce and thrive. Our nation's wildlife is a precious public resource; and enhancing and maintaining high-quality habitat on these lands are, therefore, important to all of us.

If you are such a landowner, you are to be commended for your desire to enhance the wildlife habitat on your land. If, as we hope, you put the principles you learn into practice, then all of society will owe you a vote of thanks.

The recommendations in this book were developed by incorporating the scientific literature on the subject with the broad-based experience of the authors and reviewers. Most of the examples in the text are from New England, but the principles and practices should be applicable throughout the Appalachian and northern hardwood regions of the United States.

The book is meant to be exactly what the title states: a practical guide. If we are successful, after completing each chapter you will enter your woodlot with new insights, new ideas, and greater enthusiasm. And when you've finished the final chapter, you will possess the foundation of a new and well-focused plan for enhancing wildlife habitat in your forest.

Each chapter builds carefully on the previous chapter. The first two chapters, "Basic Forest Wildlife Ecology" and "Understanding Wildlife Habitats," provide fundamental ecological and management principles. The remaining chapters describe specific habitat-enhancement techniques that incorporate these principles and the life-cycle needs of the wildlife species in question. Throughout the text you will encounter key terms in ***bold italics.*** These bold, italicized words and terms are defined in the glossary for future reference.

Included at the end of each chapter are field exercises. These exercises are an integral part of the book and are designed to help you develop first-draft habitat maps and enhancement plans for your forest. We strongly encourage you to work through them. If the topic of habitat enhancement interests you enough to read the text, you should find the exercises to be both enjoyable and rewarding.

Finally, we offer one caution: this guide is *not* intended to take the place of professional forestry and wildlife assistance in developing a habitat-enhancement plan. It should, however, help to make you a more knowledgeable landowner whose goals, objectives, and needs are clear. This, in turn, will enable you to make the most efficient use possible of the time you do have, on your land, with these professionals.

Basic Forest Wildlife Ecology

Photo: J. S. Barclay

Gaining an understanding of plant and animal interactions in the forest is the primary purpose of this first chapter. Animals interact with each other and with other living and nonliving components of forests in a complex association known as the forest ecosystem.

Introduction

When we walk through the woods, our attention may be diverted by the stunning brilliance of a wildflower, the peculiar appearance of a fungus, the rustling of a mouse scurrying through dried leaves, or the rapid-fire pounding of a woodpecker. Even if we have been preoccupied, we often become increasingly aware of our surroundings and begin to notice the complexity within forests. We may pause to wonder how all the plants and animal components of forests interact and what role wildlife plays in the total scheme of forest interactions.

Gaining an understanding of plant and animal interactions in the forest is the primary purpose of this first chapter. Animals interact with each other and with other living and nonliving components of forests in a complex association known as the forest ***ecosystem***. Understanding forest ecosystems helps us comprehend the role of wildlife in these interactions and permits us to work more effectively with resource specialists to manage wildlife within our forests.

A glossary and list of scientific names for plants and animals mentioned in the text are found in the appendix to this publication.

The Growth of Forests

Learning about the natural progression of change that occurs as our northeastern forests develop and grow enables us to understand the ways in which forests and wildlife interact. Many wildlife **species** live exclusively in forests and can be greatly affected by changes that occur within the forest. Included in this chapter is a section specifically on forests, because it is there that we will apply our strategies to encourage wildlife.

Forest Succession

Many of us are all too familiar with the little seedlings that so predictably work their way up through areas where they are most unwanted—flower beds, vegetable gardens, and mulched areas. During our kinder moments, we simply refer to them as weeds, as indeed they often are. They are merely performing their job, however, by fulfilling their role in *forest succession*.

Forest succession is the natural progression and replacement of plant and animal species, and their influences, over time. If our gardens were to be unattended, grasses and other **herbaceous plants** (nonwoody vegetation) would seed themselves in along with trees and shrubs (woody vegetation). Forests are continually changing as trees grow and die. Some changes, such as the growth of an oak to maturity, may proceed relatively unnoticed, while other changes can occur

Forest succession is the natural progression and replacement of plant and animal species, and their influences, over time.

Figure 1. Only shade-tolerant trees and shrubs can thrive in the understory of a dense forest.

rapidly, with dramatic results. Fires and logging operations, for example, can reduce dense forests to **clearings**, and hurricanes and severe ice storms can snap many tree tops and branches.

Where clearings or openings have been created by natural or human disturbances, full sunlight reaches the floor. The first herbaceous plants, shrubs, and trees that usually appear in these sun-filled openings are called ***pioneer species.*** Eastern red cedar, quaking aspen, grey birch, and pin cherry are some common examples. These species grow rapidly in full sunlight but are ***shade-intolerant***.

Shortly after these pioneers become established, a close examination will usually reveal that some additional species that can tolerate more shade begin to gain a foothold on the site. These ***intermediate species***, such as oaks and hickories, can persist in the light shade cast by the pioneers. (See table 1, page 13.) Because they are much longer lived, they eventually overtop the pioneers, which quickly begin to decline and die from the lack of sunlight.

This intermediate stage of forest succession creates a much denser canopy, and can persist for well over a century. Gradually, however, an ***understory*** of ***shade-tolerant*** species, such as sugar maple, eastern hemlock, and/or American beech begins to develop (figure 1). These species are relatively slow growing but outcompete other less tolerant species in the heavy shade. Eventually, as the ***overstory*** trees begin to age and decline, they emerge to form the new forest overstory.

These shade-tolerant species are the only ones that can reproduce under their own shade and in the absence of any disturbance could theoretically occupy a site indefinitely as the ***climax forest***. In reality, however, this virtually never occurs. Fire, hurricanes, insect infestations, diseases, and/or people with chain saws intervene, producing a disturbance and causing the forest to revert to an earlier stage of forest succession.

Succession is an important factor determining where given wildlife species are found in our forests. Forest wildlife adapts to forests in specific stages of succession (figure 2). Some wildlife species may prefer a single successional stage, while others require several stages. **The reliance of wildlife species on varying forest conditions allows us to make generalizations about managing our woodlands for wildlife.** If we wish to attract wildlife species that prefer specific successional stages, then we can manipulate our forests to contain those desired stages.

Some wildlife species may prefer a single successional stage, while others require several stages.

Special Features of Forest Succession

When forest succession begins, many of the first plants germinate from seeds which possess a variety of remarkable dispersal mechanisms. These seeds have adaptations that increase their chances of being carried great distances from the parent plant. Many are light in weight and have wings (maple, birch, pine, and ash seeds) or down (milkweed, cattail, willow, and aspen seeds) which act like helicopters or sails to carry the seed in the wind, sometimes to distant places such as the center of fields (figure 3). The seeds in berries such as cherries and elderberries are often eaten by birds and mammals and deposited at new locations in the animals' droppings. Large palatable seeds, such as nuts, may be buried and hidden by animals.

Some seeds with sticky or burr-like adaptations are transported when they attach to the fur (or clothing) of unsuspecting passersby. Jewelweed and witch hazel seeds are forcefully ejected from the parent plant when ripe, often landing several feet from their "launching pad." Buoyant seeds may be carried downstream or across water bodies by currents and wind. Following dispersal, many seeds, such as the raspberry's, remain dormant for years until conditions are favorable for germination.

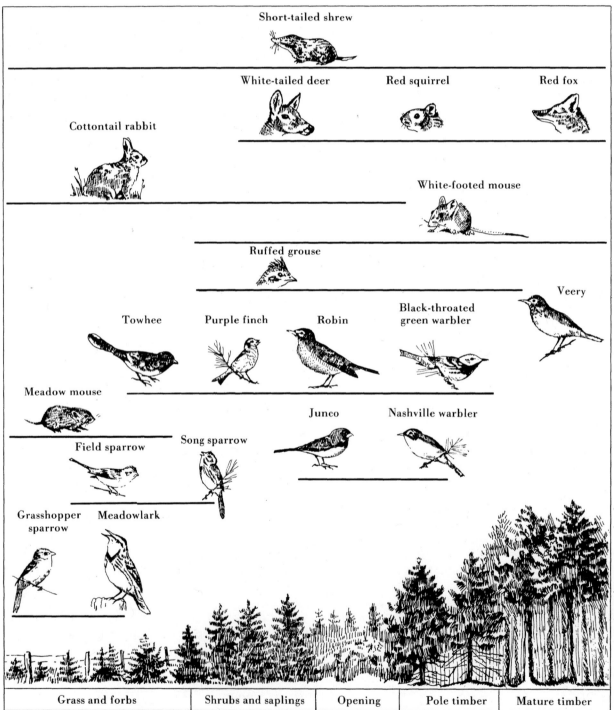

Figure 2. Each stage of forest succession is important to some wildlife species.

Plants establish themselves by a number of other means as well. Some plants, such as grasses, sumacs, and aspens, have complex root networks capable of sprouting new growth. Trees and shrubs often sprout from stumps following cutting or injury. Such sprouts have the potential to grow faster than a plant that started from a seed because a mature root system has already been established. Sprouts may also develop when branches of certain plants (such as willows) touch moist soil and establish their own root system.

The Forest Community

A ***community*** is the living portion of an ecosystem. Trees, wildflowers, fungi, mice, and woodpeckers are all members of the forest community. The area occupied by a community can vary in size. We can describe a community within a particular portion of our property, or we can expand the frame of reference to encompass a large geographic area.

The northeastern United States's plant community differs in species composition from north to south and, to some extent, from east to west. Much of northern New England is characterized by a forest type consisting of northern **hardwoods** such as beech, birch, and maple and **conifers** such as spruce and fir. The southernmost portion of New England is typically forested with hardwoods such as oak and hickory and conifers such as pine and hemlock (figure 4).

Just as communities change from location to location, they also change with time. Local plant communities are drastically affected during forest succession as some species grow and predominate while others are reduced or eliminated. Animal communities also change as the composition of plant communities changes. A brushy clearing suitable for cottontail rabbits, for example, may, in ten years, become too overgrown for cottontails but be preferred as a feeding area by ruffed grouse.

Just as communities change from location to location, they also change with time.

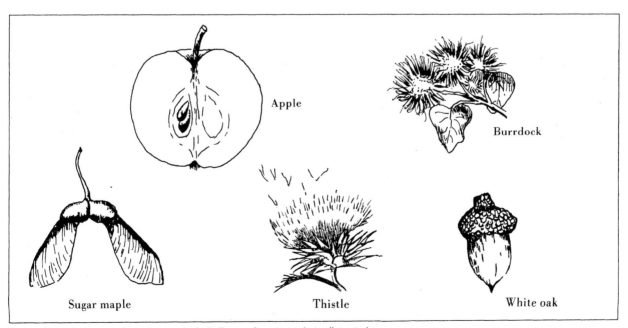

Figure 3. Birds, mammals, water, and wind all carry forest seeds to distant places.

Chapter 1: Basic Forest Wildlife Ecology

The Food Pyramid

The brilliant wildflower that caught our attention, the oak that the woodpecker fed upon, and the maple that we cut for firewood all possess qualities unique among plants. They manufacture their own food from nutrients and sunlight by the process of **photosynthesis**. Birds, mammals, and other animals do not possess this ability to provide their own food for energy. Instead, they must rely on plants or other animals for food. (See table 2, page 14.)

Plants and animals can be organized into an imaginary pyramid of **trophic**, or feeding, **levels** through which energy flows (figure 5). If we were to place all of the green plants from a given forest community on one side of a scale, and all of the animals from the same community on the opposite side, then we would see that the plants outweigh the animals by far. Green plants occupy the bottom level of the food pyramid. They represent the largest level of the pyramid and contain more stored energy than the other levels.

The second level of the food pyramid is occupied by animals which feed on plants (**herbivores** such as caterpillars, rabbits, and deer). The remaining levels of the pyramid become progressively smaller and represent animals that do not usually feed on plants (**carnivores** such as hawks, owls, weasels, bobcats, and wolves) but survive by consuming animals from lower levels. However, not all animals fit neatly into a single level. Some animals (**omnivores** such as opossums, foxes, and humans) occupy more than one trophic level because they readily select a diet of both plants and animals.

If we were to place all of the green plants from a given forest community on one side of a scale, and all of the animals from the same community on the opposite side, then we would see that the plants outweigh the animals by far.

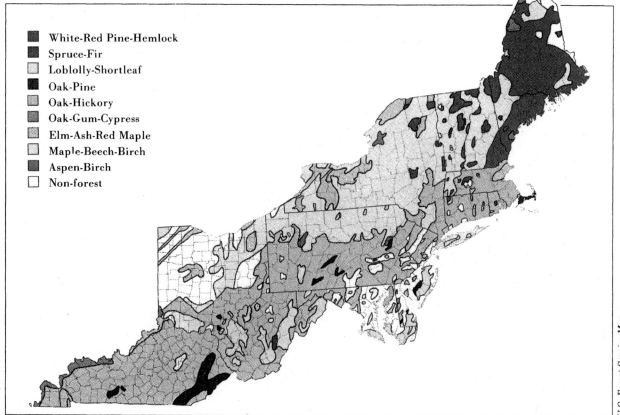

Figure 4. Forest types and the wildlife that live in them vary across the northeast region of the United States.

*U.S. Forest Service Map
Northeast Forest Experiment Station*

The transfer of energy from one trophic level to the next is relatively inefficient. Only a small amount of all sunlight that reaches the forest is utilized by green plants and other members of the forest community. The remainder either radiates as heat or reflects off vegetation, water, or the forest floor. Similarly, only a small amount of available plant energy is utilized by herbivores. An energy loss of approximately 90% usually occurs at each successive trophic level.

Seldom are more than four or five trophic levels present within an ecosystem. At the highest existing level, not enough energy is available to support a higher level. For example, it would not be feasible for an animal to hunt and feed on wolves. In order to do so, this animal would have to be larger than a wolf and capable of traveling considerable distances in search of an animal that is relatively scarce and extremely mobile (Arms and Camp 1982).

This pyramid is simply a way to envision the basic energy structure of a forest community. The number of trophic levels and the size of each level are not permanent, but fluctuate.

Food Chains and Food Webs

A *food chain* describes the progression of feeding events between plants and animals. If we were to examine the foods eaten by animals in our forest, then we could diagrammatically depict a food chain beginning with green plants and "ending" at the highest trophic level, carnivores. Links in the chain, or types of food eaten, represent different trophic levels. Unlike a food pyramid, a food chain specifically demonstrates the flow of energy from species to species. For example, the leaf of a northern red oak may be eaten by a gypsy moth caterpillar, which may be eaten by a northern oriole, which in turn may be consumed by a sharp-shinned hawk (figure 6). Or perhaps a gray fox

The leaf of a northern red oak may be eaten by a gypsy moth caterpillar, which may be eaten by a northern oriole, which in turn may be consumed by a sharp-shinned hawk.

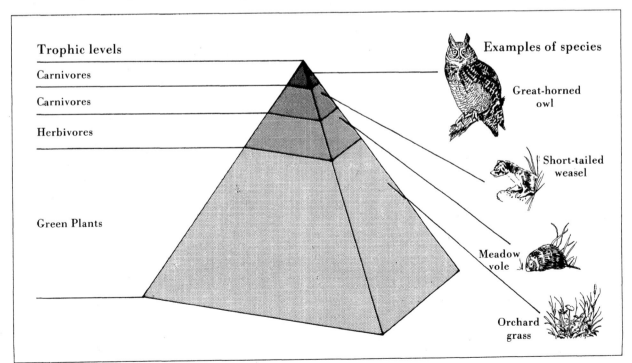

Figure 5. The number of species declines rapidly with each step up the food pyramid.

may capture a cottontail rabbit nibbling on red clover. The possibilities are nearly infinite.

The combination of various food chains forms a complex system known as a ***food web***. Each food chain can be visualized as a strand of this intricate web "spun" by the forest community. Unlike food chains, food webs represent the overall feeding patterns of animal species. Sharp-shinned hawks do not eat northern orioles exclusively but also consume other songbirds and, occasionally, small mammals.

The Role of Decomposers

When a plant or animal dies, it is broken down (decomposed) by organisms termed decomposers or ***saprophages***. Decomposition allows nutrients to be recycled within the forest ecosystem, passing from soil to plant, to herbivore, to carnivore, back to the soil, and again to a plant at the beginning of a food chain. In the absence of decomposers, nutrients would not be recycled but would all be trapped in dead plants and animals within a few years.

Of all the foliage produced annually by ***deciduous*** trees, it is unusual for more than 10% to be consumed by herbivores. Leaves that are not consumed fall to the forest floor in autumn, and the process of decomposition begins. Fallen leaves, or ***leaf litter***, begin to dissolve in the presence of water. Some minerals and organic compounds leach from the leaf material and return to the soil as nutrients in elemental form. Decomposers such as earthworms and millipedes reduce leaf litter to small particles which are further reduced to useable nutrients by specialized bacteria and fungi. The time required for complete decomposition to occur is called the ***transit time***. The longer the transit time, the greater the accumulation of leaf litter on the forest floor.

In the absence of decomposers, nutrients would not be recycled but would all be trapped in dead plants and animals within a few years.

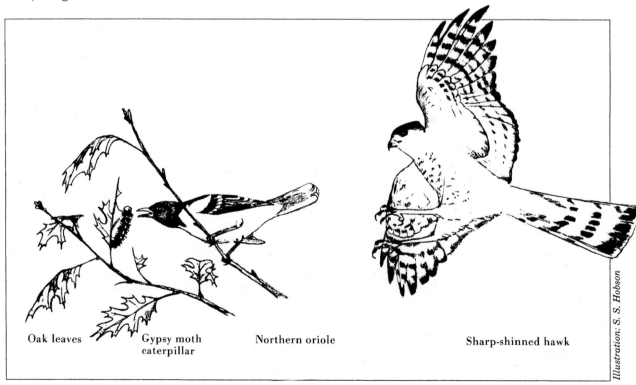

Figure 6. Oak leaves, gypsy moth caterpillars, northern orioles, and sharp-shinned hawks are all links on the same food chain.

Niche

The particular role played by each species within the forest community is called a ***niche***. We can easily conceptualize a niche if we think of it as the "occupation" of the species. For example, yellow-bellied sapsuckers are tree dwellers who typically feed on the sap and inner bark of trees and excavate their own nest cavities in trees.

Species that occupy similar niches and live in the same area will generally compete for food, while species occupying unique niches can live and feed relatively free from competition with other species (figure 7). Many species display obvious characteristics that enable them to survive in the niche they occupy. For example, the large, chisel-like teeth of porcupines are well suited for gnawing the bark of trees, and the slim, elongated bill of hummingbirds is ideal for extracting nectar from flowers (figure 8). Porcupines and hummingbirds, like yellow-bellied sapsuckers, are examples of species utilizing relatively specialized niches in which a limited number of specific foods are eaten. Some species occupy broader niches and feed on a greater variety of foods. Raccoons, red foxes, opossums, and other omnivores are examples of species occupying broad feeding niches, because they readily consume many kinds of plants and animals.

We can easily conceptualize a niche if we think of it as the "occupation" of the species.

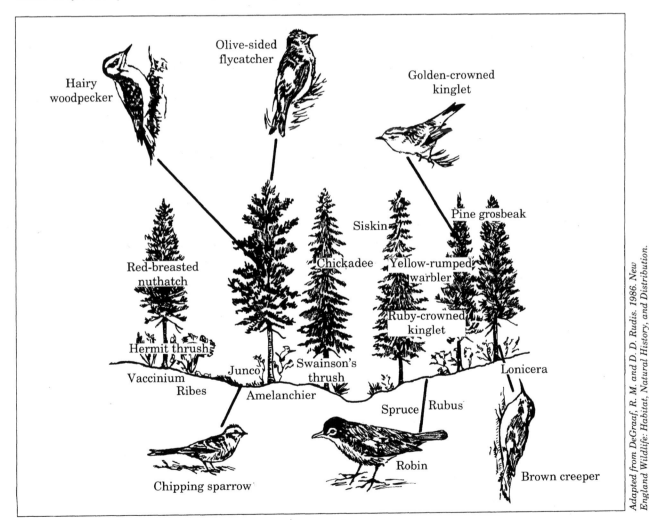

Figure 7. Every species has its special role, or niche, in the forest environment.

Species that occupy specialized niches tend to be more sensitive to changes in the forest environment than species that occupy broader niches. For example, if an isolated patch of wildflowers is visited daily by hummingbirds and is then bulldozed or becomes shaded during the process of forest succession, hummingbirds would be displaced from that site. The same event, however, would probably have less impact upon red foxes or other species which occupy niches broader than the niche occupied by hummingbirds.

Wildlife Populations

A wildlife population is a group of animals of the same species that lives within a defined area during a particular time. It is difficult to observe entire wildlife populations. Usually we are able to observe only individual animals, pairs, family groups, or flocks. The area occupied by a population may be large (several counties for black bears) or small (a 5 acre woodlot for gray squirrels), and the number of individuals contained may be relatively great (as with gray squirrels), or relatively few (as with New England's black bear population).

A given population has a density which is defined as the number of individuals per unit of land area. For example, a particular forested area might contain two gray squirrels per acre or two black bears per 70 square miles. The density of most populations is not constant throughout forests, but varies according to the capability of a given area to support individuals.

The density of most populations is not constant throughout forests, but varies according to the capability of a given area to support individuals.

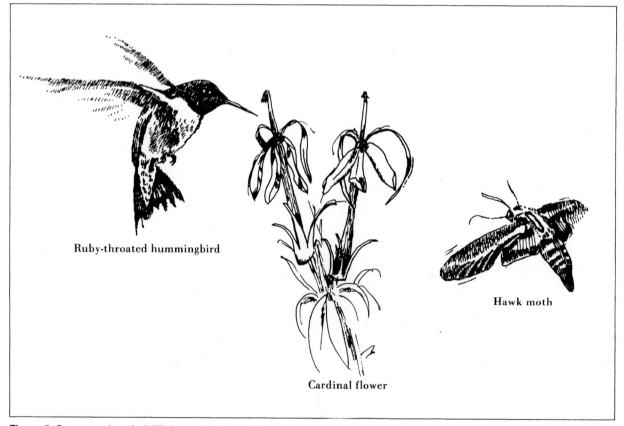

Figure 8. Some species of wildlife have developed niches which are specialized and unique.

Another concept of populations, known as age structure, refers to the number of individuals in various *age classes* of each sex. If for some reason a population consists primarily of older individuals reproducing at low levels, then this population may be in danger of disappearing. On the other hand, populations comprised chiefly of young individuals have the potential to rapidly increase in size. Usually, a healthy population includes a substantial number of members from several age classes with a sufficient number of young surviving to maturity to ensure future reproduction.

Carrying Capacity

Carrying capacity is the maximum number of individuals that an ecosystem can support over time. Stated differently, carrying capacity is the optimal population density from year to year for a species living on a particular piece of land. When a population is at or below carrying capacity, each individual has sufficient food and space to survive. A 40-acre forest, for example, might be capable of supporting ten ruffed grouse. When ten ruffed grouse occupy this area, there will be potential for different species to occupy different niches, but not for additional grouse. The addition of individuals to a population already at carrying capacity causes crowding and a strain on available food sources and other essential wildlife requirements.

Carrying capacities can change due to changing conditions within the forest. The production of food crops often varies annually and can change the carrying capacity over successive years. Surplus acorn production, for example, can temporarily increase carrying capacities for gray squirrels. Changes in carrying capacity often cause a temporary imbalance of individuals until the population adjusts up or down to the new carrying capacity.

When a population is at or below carrying capacity, each individual has sufficient food and space to survive.

Reproduction

The frequency of birth or hatching and the number of young produced differs among wildlife species. Some species (especially those near the bottom of the food pyramid) produce large numbers of young several times a year, while others (carnivores at the top of the food pyramid) produce fewer young once a year. A species that successfully reproduces once a year usually cannot be encouraged to raise additional young in the same year. Quality living conditions and ample food resources, however, can encourage them to produce more offspring in a single litter or clutch.

Wildlife species do not mass produce. Even desirable forestland will sustain only a certain number of individuals of a population (based on the limits of carrying capacity). Excessive reproduction can temporarily lead to overcrowding and result in detrimental conditions such as malnutrition, if not excessive mortality. Overcrowding in deer herds, for instance, often leads to over-browsing and disruption of healthy interactions between plant and animal. Areas may become so over-browsed that the forest can no longer support more deer, new plant growth is hindered, and carrying capacities for other wildlife populations are reduced.

Mortality

It can be disheartening to find a deer carcass or the remains of other wildlife species that we work diligently to manage. But, by nature, each living individual of the forest ecosystem will die—wildlife is no exception. Mortality results from many causes, such as accidents, severe weather, predation, disease, or (rarely in the wild) old age.

Deaths due to accidents are relatively uncommon but may occur if an animal breaks a wing or leg, for example. The addition of manmade structures to the environment can increase the frequency of wildlife accidents. Radio towers and large panes of window glass often lead to numerous bird collisions, and certain utility wires have caused electrocution when birds or mammals span two wires. Motor vehicles also account for many deaths.

Extremes in temperature, wind, and precipitation can result in many losses. Small birds, such as chickadees, can die during bitter cold nights if their energy reserves become depleted before the dawn feeding period. Strong winds, especially if combined with cold temperatures and precipitation, can force animals to remain within protective shelter for prolonged periods of time. Faced with such conditions, they must either confront the elements and search for food or face the threat of starvation. Prolonged spring rains coupled with cool temperatures can cause heavy losses when the broods of ground-nesting birds, such as woodcock, ruffed grouse, and wild turkeys, die of exposure after becoming damp and chilled.

Predation accounts for a significant amount of wildlife mortality. Densities of **predator** populations depend largely on the availability of their **prey**. When a prey species becomes numerous, predators may also become more numerous. We may notice more predators as our property becomes more attractive to the species we manage. Increased predator sightings are often good indications that our management efforts are effective.

Fatal diseases can result from nutritional deficiencies, infections, or toxic chemicals (Bailey 1984). Animals living in poor-quality forests are often more susceptible to diseases than animals living in areas containing adequate supplies of nutritionally balanced food resources.

Wildlife species with high reproductive rates, such as birds and small mammals near the bottom of the food pyramid, typically exhibit high mortality rates. They produce more offspring than their immediate environment can support, so individuals die by some means until the carrying capacity is achieved. In populations of such species, the death of surplus individuals is known as **compensatory mortality**.

For example, one mallard pair might produce a clutch of nine eggs in an area that is capable of supporting an average of four mallards—two parents and two offspring to replace the parents. Two eggs might become chilled by a cold rain and not hatch. Of the seven eggs that hatch, one duckling might be eaten by a water snake, two might fall prey to red foxes, another to a Cooper's hawk, and the fifth might succumb to disease. Each mortality-causing factor is an example of compensatory mortality. If one factor had not led to the death of a duckling, then another factor probably would have. In the absence of such mortality, our hypothetical pair of mallards and their offspring theoretically could produce 13,995 individuals in five years.

We may notice more predators as our property becomes more attractive to the species we manage. Increased predator sightings are often good indications that our management efforts are effective.

Table 1. Characteristics of common northeastern trees and woody shrubs.

	Moisture Regime*	Root System	
Trees			**Pioneer Species**
Quaking aspen (Populus tremuloides)	5	shallow and spreading	– intolerant of shade, need full sun
Bigtooth aspen (Populus grandidentata)	2–3	moderate and spreading	
Eastern red cedar (Juniperus virginiana)	1–3	deep	
Pin or fire cherry (Prunus pennsylvanica)	1–3	shallow	
Sassafras (Sassafras albidum)	1–3	moderate	
Black locust (Robinia pseudoacacia)	5	adaptable	
Yellow poplar (Liriodendron tulipifera)	2–3	deep and spreading	
Black willow (Salix nigra)	4	shallow	
Gray birch (Betula populifolia)	1–3	shallow	
Woody Shrubs			
Autumn olive (Elaeagnus umbellata)***	5		
Staghorn sumac (Rhus typhina)	1–3		
Multiflora rose (Rosa multiflora)***	5		
Common juniper (Juniperus communis)	1–3		
Speckled alder (Alnus rugosa)	3–4		
Trees			**Intermediate Species**
Northern red oak (Quercus rubra)	2–3	moderate to deep	– some tolerance of shade, need mostly sun to partial sun
Scarlet oak (Quercus coccinea)	1–2	deep	
White oak (Quercus alba)	1–3	deep	
Black oak (Quercus velutina)	1–2	deep	
Swamp white oak (Quercus bicolor)	3–4	moderate	
Chestnut oak (Quercus prinus)	1–2	deep and spreading	
Shagbark hickory (Carya ovata)	5	deep	
Pignut hickory (Carya glabra)	1–3	deep	
Mockernut hickory (Carya tomentosa)	5	deep	
Bitternut hickory (Carya cordiformis)	2–3	deep	
Red maple (Acer rubrum)	5	shallow but adaptable	
Yellow birch (Betula alleghaniensis)**	2–4	shallow and spreading	
Black or sweet birch (Betula lenta)**	5	moderate and spreading	
Eastern white pine (Pinus strobus)	5	deep but adaptable	
American elm (Ulmus americana)**	3–4	shallow and spreading	
Slippery elm (Ulmus rubra)	5	shallow and moderate	
White ash (Fraxinus americana)**	2–4	shallow to moderate	
Black cherry (Prunus serotina)	5	shallow	
Hop-hornbeam (Ostrya virginiana)	2–4	moderate	
Butternut (Juglans cinerea)	5	adaptable	
Basswood (Tilia americana)**	2–3	deep and spreading	
Woody Shrubs			
Viburnum (Viburnum spp.)	5		
Witch hazel (Hamamelis virginiana)	2–3		
Winterberry (Ilex verticillata)	3–4		
Mountain laurel (Kalmia latifolia)	5		
Highbush blueberry (Vaccinium corymbosum)	3–4		
Huckleberry (Gaylussacia baccata)	1–2		
Hazelnut (Corylus spp.)	5		
Elder (Sambucus spp.)	2–4		
Trees			**Climax Species**
Sugar maple (Acer saccharum)	2–4	adaptable	– tolerant of shade
American beech (Fagus grandifolia)	2–4	shallow and spreading	
Eastern hemlock (Tsuga canadensis)	5	shallow and spreading	
American hornbeam (Carpinus caroliniana)	3–4	shallow	
Woody Shrubs			
Spicebush (Lindera benzoin)	3–4		
Pepperbush (Clethra alnifolia)	3–4		

* Moisture regime where commonly found:
  1. dry   2. well drained   3. moderately drained   4. poorly drained to wet   5. wide range
** More tolerant than others in this category; often associated with climax species
*** Not native species

Table 2. Relative food values of selected northeastern trees and shrubs for wildlife [1] and for honey bees [2] by stage of forest development.

	Overall Rating [3]	Water Birds	Marsh/Shore Birds	Game Birds	Song Birds	Fur/Game Mammals	Small Mammals	Browsers	Honey Bees	
Trees										
Aspen	M	O	O	M	L	H	L	M	O	**Pioneer**
Cedar	M	O	O	O	H	L	O	O	O	**Species**
Sassafras	L	O	O	L	H	L	O	L	O	
Black locust	L	–	–	–	–	–	–	–	–	
Yellow poplar	M	O	O	O	H	L	L	L	H	
Willow	M	O	O	L	O	H	L	M	M	
Birch	H	O	O	M	H	H	L	M	H	
Cherry	H	O	O	L	H	M	L	L	M	
Shrubs										
Autumn olive	M	–	–	–	–	–	–	–	–	
Sumac	H	O	O	M	H	H	O	L	H	
Wild rose	M	–	–	–	–	–	–	–	–	
Juniper	H	–	–	–	–	–	–	–	O	
Speckled alder	M	O	O	L	M	M	O	L	L	
Trees										
Oaks	H	M	L	H	H	H	M	L	L	**Intermediate**
Hickory	M	O	O	L	H	H	L	O	L	**Species**
Maple	H	O	O	M	H	H	M	M	H	
Birch	M	–	–	–	–	–	–	–	–	
Pine	H	O	O	L	H	H	L	L	O	
Elm	M	L	O	L	H	M	O	L	L	
Ash	M	L	O	L	H	M	L	L	O	
Hop-hornbeam	L	–	–	–	–	–	–	–	–	
Butternut	L	–	–	–	–	–	–	–	–	
Basswood	M	–	–	–	–	–	–	–	H	
Shrubs										
Viburnum	L	–	–	–	–	–	–	–	L	
Witch hazel	L	–	–	–	–	–	–	–	O	
Winterberry	L	–	–	–	–	–	–	–	O	
Laurel	L	–	–	–	–	–	–	–	M	
Blueberry	H	L	O	L	H	H	M	L	M	
Huckleberry	L	–	–	–	–	–	–	–	M	
Hazelnut	M	O	O	M	L	H	M	L	L	
Elder	M	O	O	L	H	M	L	L	O	
Trees										
Maple	H	–	–	–	–	–	–	–	–	**Climax**
Beech	H	L	O	L	H	H	L	L	O	**Species**
Hemlock	M	O	O	L	H	M	L	L	O	
Hornbeam	L	–	–	–	–	–	–	–	O	
Shrubs										
Spicebush	L	O	O	L	H	O	O	O	M	
Pepperbush	L	–	–	–	–	–	–	–	–	
Total Plant Taxa										
H = 10		0	0	1	16	11	0	0	5	
M = 14		1	0	5	1	5	4	4	5	
L = 9		4	1	12	2	3	10	13	6	
O = 0		15	19	2	1	1	6	4	12	

1. Based on the list in table 1 and analyses of regional data in Martin, Zim & Nelson, 1951. *American wildlife and plants.*
2. Based on results of 1988 survey and supplemental sources (Barclay, unpubl.) which provided data that help characterize woody plant values to bees and to wildlife.
3. H = high; M = moderate, L = low; O = negligible or none

Review Questions

1. What is forest succession? Why is this an important concept in wildlife management?

2. What is carrying capacity? Why is this an important concept in wildlife management?

3. What is a food pyramid? Why is each layer in the pyramid smaller than the one below?

Chapter 1: Basic Forest Wildlife Ecology

Field Exercises

1. Using table 1 on page 13 (tree species found in each successional stage) and, if helpful, a tree identification guide, walk through a portion of your woodlot and see if you can answer the following questions.

    What stage of succession is the area in: pioneer, intermediate, or climax?

    Can you determine how trees might have established themselves, e.g., stump sprouts, root suckers from a parent plant, wind-blown seeds, or seeds carried by birds and mammals?

    If an intermediate or climax stage is present, can you see any evidence of an earlier stage, e.g., dead or dying red cedars in the understory of an oak/mixed hardwood forest?

    Are other successional stages present on your woodlot?

2. During the summer or fall, gather some seeds, berries, acorns, nuts, burrs, etc.

    What are the dispersal mechanisms of each?

    What do the mechanisms tell you about how far the seed inside can travel?

    Do plants in a given successional stage tend towards any particular methods of dispersal?

3. Observe three species of wildlife on your woodlot (binoculars may be helpful).

    Where do each of these species spend most of their time when feeding? For example (if you choose to observe a bird), many birds eat insects, but does the species that you identify scratch through the leaf litter, search in the lower branches of young trees and shrubs, stay in the forest canopy, or creep along tree trunks and large branches?

    If you can see what each species is eating, sketch a food chain and guess what links are missing, if any, to complete a food chain for your woodlot.

    Based on your observations, describe the niche, or "occupation," of each species.

4. With the assistance of your state service forester or a local consulting forester, make a map of your woodlot. Delineate vegetation **cover types** (e.g., fields, wetland, forests, and the primary tree species present in each forested section); and include woods roads, trails, a north arrow, an approximate scale, and the date. Retain a master copy and make a photocopy for use in chapter 2.

These exercises were designed to be done on your own woodland. Once completed, you will be well on your way toward your own wildlife management plan.

2 Understanding Wildlife Habitats

We cannot force wildlife species to live in a particular place, so we must concentrate on manipulating habitats to accomplish our management goals of attracting them.

Introduction

To effectively manage our forests for wildlife, we should become familiar with the elements that determine which areas wildlife choose to inhabit. A wildlife **habitat** is any area that contains all requirements essential to the survival of wildlife. Stated simply, it is the area in which any given animal species lives. We cannot force wildlife species to live in a particular place, so we must concentrate on manipulating habitats to accomplish our management goals of attracting them.

Habitat Requirements of Wildlife

Wildlife habitat is composed of three main elements that we can work with to encourage desired species: **cover** (shelter), food, and water. These elements must be available in proper proportions and located within appropriate distances of one another. The quantity, quality, distribution, and seasonal availability of these three elements are important in determining the carrying capacity, or the number of individuals a habitat can support over a given period of time. This number is not constant, but may change as certain habitat elements change. Physical space is a fourth habitat element and is also affected by the degree to which cover, food, and water are available.

When an element is in short supply for a population of animals in an area, that element is called a ***limiting factor***. Often it is possible to increase the number of individuals in a population by identifying limiting factors and managing habitats to provide more of the element in short supply. For example, if adequate cover exists but food is insufficient, food sources, the limiting factor, can be increased. Generally, as the constraints of limiting factors are eased, habitat quality is improved, and the carrying capacity of an area for a given species will increase.

Cover

"Cover" is the protective element within an animal's habitat and is a factor in one or more of an animal's necessary life functions, i.e., breeding, nesting, hiding, resting, sleeping, feeding, and traveling. Cover may be a hemlock tree for a screech owl (sleeping cover), a stonewall for a chipmunk (escape cover), or a dense patch of brush for a deer (resting cover) (figure 9).

Cover provides physical structure to wildlife habitats. Vegetation fulfills the cover requirement of most wildlife species, but ledges, rock slides, caves, burrows, and man-made structures are also utilized. When considering manipulation of vegetation to provide cover for wildlife, it is important to know the cover requirements of the particular species of concern. With this information we can determine which plant species are appropriate to encourage and how they should be distributed in an area.

When considering manipulation of vegetation to provide cover for wildlife, it is important to know the cover requirements of the particular species of concern.

Figure 9. All wildlife need some form of protective cover.

An animal's cover requirements are variable. For example, deer and grouse generally feed in relatively open areas of forest, but during a midwinter storm with freezing rain and winds, they may seek refuge in a dense stand of conifers, such as fir, hemlock, pine, or spruce. The need for cover may change with the seasons, weather, prevalence of predators and pests, or behavioral characteristics between different ages and sexes, as in nesting birds.

Vegetative cover is not static, but continually grows in a somewhat predictable process of forest succession. Saplings and shrubs that currently serve as ideal cover for brown thrashers will grow and become poor thrasher cover in several years if left unmanaged. The saplings will shade out the shrubbery, making the birds' cover less available and eventually nonexistent. If we wish to attract certain wildlife species, then we must recognize the process of forest succession and manage vegetation accordingly.

Special Examples of Cover

Caves and Ledges

Both "caves" (i.e., rock shelters, crevices, and caverns) and ledges are important examples of non-vegetative cover (figure 10). Caves provide shelter for wildlife such as bears, bobcats, porcupines, raccoons, and bats. Ledges are often found in hilly areas with rock outcrops. They may be used as nest sites by birds such as turkey vultures, cliff swallows, and ravens, and may be occupied by bats where ledge walls extend far enough inward.

The need for cover may change with the seasons, weather, prevalence of predators and pests, or behavioral characteristics between different ages and sexes, as in nesting birds.

Figure 10. Rock outcrops are an additional source of shelter for many animals.

Cavity Trees

Natural **cavities** develop when part of a tree dies or is injured. Death or injury can result from fire, insect attack, wind, logging wounds, herbicides, snow or ice storms, and other causes (DeGraaf and Shigo 1985). Organisms such as bacteria, fungi, insects, and **nematodes** may become established in the tree wounds and slowly digest the affected wood until a cavity results (figure 11). Woodpeckers, which are primarily excavators, create cavities when they excavate holes for feeding (figure 12), nesting, and sleeping (roosting). The resulting cavities are used extensively by many species of mammals and birds.

Snags

Snags are dead, standing trees. They provide nesting and sleeping cover for wildlife and are equally important for supplying many birds with a diet of insect larvae. Snags often contain cavities formed by previous injury, heavy decay, and woodpeckers (figure 13).

Dense Vegetation

Although areas of thick growth are usually avoided by the casual passerby, they are valuable to wildlife for protection. A tangled mass of vines and brambles provides escape cover for rabbits, and dense laurel thickets or conifer stands provide crucial winter cover for deer and grouse.

Although areas of thick growth are usually avoided by the casual passerby, they are valuable to wildlife for protection.

Figure 11. Tree cavities provide protective cover for many wildlife species.

Figure 12. The woodpecker creates cavities as it feeds.

Food

To maintain good health for growth and reproduction, animals must have an adequate supply of nutritious foods. If a site supports fewer than normal native species, the reason may be attributed to a deficiency in food quantity and/or quality.

The presence of food does not always guarantee the presence of wildlife, however. Other habitat elements may be in short supply (limiting factors). In some cases, food may be located too far away from cover to benefit the species of interest. Food plants, however, can serve as both cover and a food source. Such is the case with aspen for ruffed grouse and oak for gray squirrels.

At times, food may be present but inaccessible. Black cherries may dangle from a twig too thin to support an opossum, or forage may extend just beyond reach of a deer feeding in an over-browsed habitat. The seasons and weather can be important factors determining food availability. Insects and mast crops (nuts and fruits) are only seasonally available. Severe freezing rains can coat food plants with an unbreakable layer of ice, making it difficult or impossible for species such as ruffed grouse to balance on twigs and eat buds.

As is the case with cover, requirements for food vary. Seasons, weather, and differences between species, sexes, ages, and seasonal

If a site supports fewer than normal native species, the reason may be attributed to a deficiency in food quantity and/or quality.

Figure 13. Dead, standing trees called snags contribute food, cover, and diversity to the forest environment.

Chapter 2: Understanding Wildlife Habitats

functions such as hibernation, migration, and pregnancy determine what foods wildlife need and prefer. For example, although herbaceous woodland openings are less essential to adult turkeys and grouse, young individuals need such openings to supply an abundance of high protein insects.

Through millions of years of evolution, wildlife species have adopted specific feeding niches. In other words, certain foods are generally eaten within particular habitats. For example, woodcock and robins both consume earthworms. Although their feeding locations occasionally overlap, each species usually feeds within its respective habitat—woodcock in alder swales and robins in clearings and lawns.

Manipulation of the food element of a wildlife habitat requires knowledge of the food preferences of desired species. Knowledge of the fruiting habits, persistence, period of availability, soil/site requirements, and other characteristics of food plants is helpful also. Careful selection of food plants for their additional cover qualities can increase their overall benefit to the species. Selection can also be made to provide a diversity of food types. For example, plants that mature at different times, or those that retain their fruits well into winter, can be introduced onto a site.

Mast is the seed and fruit of a tree or shrub that is eaten by animals.

Special Examples of Food Sources

Mast

Mast is the seed and fruit of a tree or shrub that is eaten by animals. It can be either hard or soft. Nut trees such as oaks, hickories, walnuts, and beeches produce hard mast. Soft mast includes catkins, berries, or other fruits from plants such as birches, dogwoods, viburnums, blueberries, and hawthorns (figure 14).

Figure 14. Maintaining a variety of native mast-producing trees and shrubs is an important part of wildlife management.

Mast is abundant periodically and, when available, is used exclusively by some wildlife species. It is also an important fall and winter food source for many species and can contribute substantially to the year-round diet. (See table 3, page 33.) Various types of mast are often the primary high-energy foods available to forest wildlife during winter. Acorns are especially important because of their large contribution to the total forest wildlife food base. Beechnuts can also contribute significantly to the total mast crop. Some wildlife species store mast to ensure that future supplies will be available.

Knowledge of the fruiting patterns of mast-producing trees helps in making decisions concerning wildlife management. Fruiting habits vary by tree species and locality and among individual trees. The fruiting habits of the white oak group, for example, differ from those of the red oak group. White oaks flower and bear fruit in one growing season, and their acorns are found on the current year's growth. The red oaks also flower and set fruit in one growing season, but their acorns are not mature until the following season. Consequently, a mast failure of all oaks in the same year is rare. Therefore, it is wise to manage for representatives of both the white and red oak groups where conditions permit.

Wolf Trees

When a tree grows in an old pasture or a forest opening without heavy competition from other trees, it usually produces a large, widespread crown with many branches. Such trees are referred to as **wolf trees** (figure 15); they are of little value for lumber or fuelwood. These trees often produce larger mast crops and can provide better nest sites than smaller trees of the same species. Recognizing and maintaining wolf trees that are good food producers can greatly benefit wildlife.

Apple Trees

Wild apple trees can be a high-preference food source for many wildlife species. Wild apple trees and old, overgrown apple orchards abound in the Northeast. Where present, salvageable apple trees should be retained and freed of competing trees by "release cutting," the removal of competing species to promote the growth of desirable species. Pruning also is helpful for encouraging fruit production.

Water

Water is essential for all wildlife. Sources of water within a wildlife habitat include dew, **succulent vegetation**, and open water. Not all wildlife species need open water, but for those requiring it, a pond, stream, or spring will serve the purpose. Required or not, most animals opt to use open water when available. Deep pools in intermittent streams can be especially important during drought periods.

Water sources should not be altered in any way that will create adverse environmental impacts. State and municipal regulations dictate how **wetlands** can be manipulated. As a general rule, streams should be protected by a buffer strip of uncut trees and other vegetation at least three times the stream width to prevent erosion, siltation, and increased water temperatures. This rule also applies where

Water sources should not be altered in any way that will create adverse environmental impacts. State and municipal regulations dictate how wetlands can be manipulated. As a general rule, streams should be protected by a buffer strip.

grazing occurs. Buffer strips should be maintained around all water resources to retard and filter surface runoff and to avoid siltation, overgrazing, and trampled vegetation on embankments.

Special Examples of Water Sites

Riparian Zones

Vegetated areas bordering free-flowing or standing water are called **riparian zones**. They vary considerably in size and species composition depending on characteristics such as gradient (slope), aspect (compass direction of slope), topography, soil type, water quality, type of stream bottom, elevation, and plant community. Riparian zones and wetlands are the most productive wildlife habitats in many areas of the Northeast. Greater availability of water to plants, frequently in combination with deeper soils, increases plant production and diversity, and generally produces more food and cover for wildlife.

Riparian zones and wetlands are the most productive wildlife habitats in many areas of the Northeast.

Spring Seeps

Spring seeps are sites where ground water becomes exposed, i.e., discharges to the surface. Although these areas sometimes remain unnoticed or are considered insignificant, they offer a valuable water

Figure 15. Wolf trees, which develop in the open without competition, are valuable for food and cover.

Figure 16. Ecotones, i.e. transition zones between different plant communities, can be especially productive wildlife habitat.

and food source for wildlife. In winter, spring seeps tend to be warmer than surface waters, increase snowmelt, and provide a variety of foods in snow-free areas for ground-feeding wildlife. If water availability is a suspected limiting factor in a given habitat, a small portion of spring seeps can be dug out by hand to form small pools that are more accessible for wildlife use.

Additional Habitat Components to Consider

Edge and Ecotone

Edges are borders where different plant communities or similar communities of different ages come together. The actual area of transition between these communities is called an ***ecotone*** (figure 16). For example, where field and forest meet, there is a border and generally an ecotone with a lush growth of weeds, shrubs, and saplings. Such areas have great plant diversity and usually sustain more wildlife

If water availability is a suspected limiting factor in a given habitat, a small portion of spring seeps can be dug out by hand to form small pools that are more accessible for wildlife use.

Figure 17. This aerial view of a variety of plant communities and land uses illustrates interspersion and edge habitats.

than either adjoining area. Wildlife species inhabiting the field or forest will occasionally frequent the ecotone, while other species will use it as their exclusive habitat. Edges and their ecotones provide wildlife with more options for cover; food; and, where edges are within riparian zones, water.

Riparian zones provide important edges, so it is desirable to protect them from harmful disturbance. Consider the areas along streams, for example. More wildlife, especially birds, will usually be noticed in these areas than in a uniform patch of adjacent woodland.

As more plant communities meet, the number of edges and micro-habitats increases; a greater **diversity** of vegetation is achieved; and more wildlife species may be attracted within limits. Therefore, where desirable, we can to some degree increase the **interspersion** of vegetation to increase the amount of edge habitat. This makes a greater number of vegetative conditions available for those wildlife species requiring more than one vegetative type (figure 17).

The size and shape of ecotones are also important (figure 18). An abrupt, narrow ecotone (A) typically has less value to wildlife than a wider ecotone of the same length (B). Narrow ecotones often have relatively little plant diversity and limited area in which to support wildlife. As ecotone size increases, the chance of attracting more wildlife species increases. Greater area can be attained by manipulating ecotone shape (C). Irregular, meandering ecotones contain larger areas, provide more visual barriers, and often hold a greater potential for sustaining wildlife than ecotones that follow straight lines or gradual curves.

Irregular, meandering ecotones contain larger areas, provide more visual barriers, and often hold a greater potential to sustain wildlife than ecotones that follow straight lines or gradual curves.

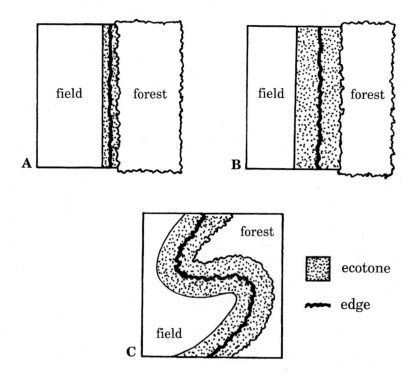

Figure 18. Possible configurations of forest vs. disturbed area ecotones on the same parcel. Option C offers the greatest combination of food, cover, and length of edge.

Wildlife Movements

Wildlife species are not restricted to one small area. We should not be concerned when we see our summer songbirds replaced by other songbird species after the first snowfall, or when we see the wild turkeys that feed on our property also feeding on our neighbors' land. Some wildlife use our woodlands seasonally; others seem to stay indefinitely. These movements to different sites are natural occurrences.

Home Range

The area that an animal occupies during its daily activities is called a ***home range***. Included within the home range are locations where the animal finds cover, food, and water. In some habitats these three requirements will be located great distances apart, forcing the animal to utilize a larger home range than if all habitat requirements were in close proximity. Some wildlife species normally occupy extensive home ranges, depending upon the nature of their food source and the availability of other habitat requirements. (See table 4, page 34.) A white-tailed deer would be expected to have a larger home range than a deer mouse, for example, due in part to its food (energy) requirements, preferences, and its larger size.

Territoriality

Many animals defend areas, or ***territories***, for their own temporary use. The area defended may be as large as the home range, but is often smaller. By claiming a territory, an animal conserves energy by lessening competition for one or more habitat resources (the elements of cover, food, water, and space). Wildlife competition occurs most readily between members of the same species or members of different species requiring similar resources.

Depending upon the wildlife species, territory occupants may be individuals, breeding pairs, or groups. Occupants often mark territories with visual cues and scents and defend them with threatening antics and displays. Such behavior alerts potential intruders and generally prevents dangerous (and costly) confrontations.

Except when they must be defended against intruders, territories enable the holders to perform their daily activities in relative solitude. For many species, territories are held only seasonally to assure that breeding and care of the young can be accomplished with little interruption and some safety. Individuals that are unable to establish territories may try to breed in unsuitable sites and are often unsuccessful at breeding attempts. Such members may suffer from poor health and wander in search of vacant territories. When a territory becomes empty (due to death or illness of the occupant, for example), it is quickly reoccupied if other individuals are available to fill it.

Some wildlife species establish territories, but only a given number of territorial animals will actually occupy a plot of land. The number of territories established will be based on the carrying capacity created by available habitat resources and on the occupants' tolerance of adjacent territory holders. As habitat quality improves, territory size usually decreases while population size increases.

Many animals defend areas, or territories, for their own temporary use. By claiming a territory, an animal conserves energy by lessening competition for one or more habitat resources.

Dispersal

You have probably heard reports of ruffed grouse flying into windows in autumn or have noticed seasonal increases in the number of mammals struck by cars. Many times, these animals are young individuals of a territorial species. As the young mature, they leave the territory in which they were raised and disperse in search of a new home range.

Such ***dispersal*** assures that parents have sufficient habitat resources within their home range. If young should return, they may be considered as intruders and possibly confronted. Dispersal divides the family and produces a more or less random scattering by which vacant or depleted areas can be reoccupied. In this way, genetic diversity is continually introduced into a given area and the frequency of ***inbreeding*** is reduced.

Migration

Migration is a seasonal movement between home ranges. It is most common in birds, but not all birds migrate. For example, woodcock, waterfowl, hawks, and many species of songbirds alternate between breeding ranges and winter ranges by migrating, but ruffed grouse, wild turkeys, and some songbirds do not. We frequently think of migrations as being strictly north-south movements of considerable distance, when in fact, they can be of variable distance and in east-west directions or between low elevations and high elevations. Migration patterns are species dependent. The arctic tern and golden plover are among the champions of long-distance migration, flying thousands of miles each season. At the other extreme, wood frogs may move only as far as the nearest spring puddle to reach their breeding grounds.

Travel Lanes

Some animals repeatedly use specific trails and runways when moving. These may be paths through open woods, within laurel thickets, or under blackberry vines (figure 19). Trails generally are chosen to afford the user some degree of protection and usually lead to food or other requirements plus nearby escape cover. If we wish to clear large sections of land or remove dense growth, we should remember that many wildlife species are reluctant to stray too far from protective cover. In these situations, we might consider leaving uncut strips, or travel lanes, so that wildlife can move more readily throughout the area. Conversely, cleared strips in mature forests can grow into brushy travel lanes for species such as deer. For many species of wildlife, edges, hedgerows, and stonewalls serve as important travel lanes, guiding and protecting simultaneously.

If we wish to clear large sections of land or remove dense growth, we should remember that many wildlife species are reluctant to stray too far from protective cover. In these situations, we might consider leaving uncut strips, or travel lanes, so that wildlife can move more readily throughout the area.

Special Example of Travel Lanes

Stonewalls (Stone Fences)

In addition to being aesthetically pleasing reminders of our heritage, stonewalls provide structure to wildlife habitats and are especially valuable as travel lanes. For small mammals, such as chipmunks, stonewalls serve as important cover for nearly all daily functions. For larger species, stonewalls provide protective cover along which to travel. Where walls border fields or woodland access roads, lush herbaceous edges may be present (or can be encouraged), further enhancing the attractiveness to wildlife.

Observing Wildlife and Recognizing Signs of Presence

It is important that we periodically check our property for wildlife use. The effort we expend depends on personal preference and can vary from a leisurely walk to a carefully planned study. We can improve our chances of observing wildlife by being at the right place at the right time. The appropriate location may be the intersection of two runways for white-tailed deer, near a den tree for squirrels, or

In addition to being aesthetically pleasing reminders of our heritage, stonewalls provide structure to wildlife habitats and are especially valuable as travel lanes.

Figure 19. Some animals, such as deer, often rely on familiar pathways (trails) from one area to another.

Chapter 2: Understanding Wildlife Habitats

beside a wooded stream or pond for raccoons. Optimal times are usually just before dusk or just after dawn, when many wildlife species are most active.

In addition to observing wildlife directly, we can observe wildlife signs to determine the extent of wildlife activity (figure 20). Often this is much easier than actually seeing the animals themselves, especially for particularly wary animals (such as foxes) and those most active at night (such as raccoons and opossums).

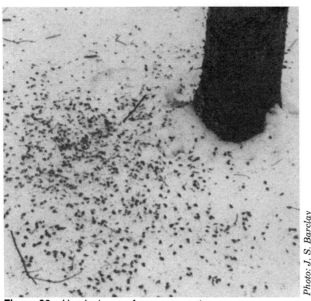

Figure 20a. Hemlock cone fragments on the snow are a feeding sign of the red squirrel.

Figure 20b. Red maple sprouts are browsed by white-tailed deer.

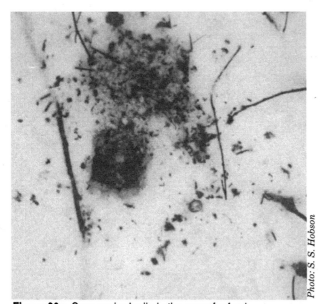

Figure 20c. Gray squirrels dig in the snow for food.

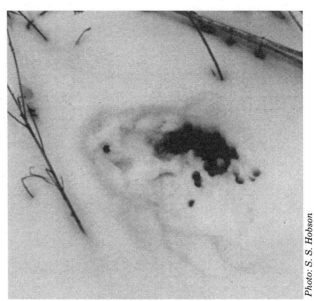

Figure 20d. The white-tailed deer leaves clumps of droppings (pellets).

Tracks

Identifying tracks, usually footprints, is an excellent means of confirming the presence of wildlife and deciphering a species' movement patterns. Those that can be followed for any distance often lead to preferred cover and favorite feeding sites. They are most easily observed in snow, moist sand, and mud. To increase the chances of detecting tracks, soft ground can be created by raking and watering likely activity areas. Two of the many field guides which can assist in track identification are listed in the "Literature Cited and Selected References" section in the back of this publication.

To increase the chances of detecting tracks, soft ground can be created by raking and watering likely activity areas.

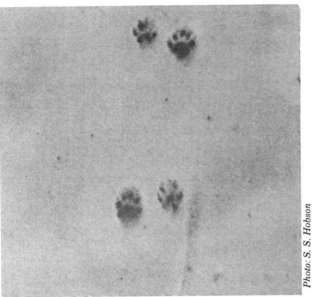

Figure 20e. Otter tracks typically appear paired as a result of the loping or bounding gait of the animal.

Figure 20f. Coyote tracks are more pointed than those of the dog or fox.

Figure 20g. Pheasant and deer have passed through the same area.

Figure 20h. Male turkey tracks show a long middle toe.

Feeding Evidence

The presence of certain species can be determined by signs of their feeding activities. Some animals, especially chipmunks, red squirrels, and gray squirrels, leave piles of seed husks and nut shells (midden piles) at sites where they feed. Deer feed on woody twigs, sprouts, and saplings, typically leaving the twig and bud torn off with a ragged appearance. Rabbits sharply "clip" twigs and chew the bark of various shrubs and tree seedlings, saplings, and sprouts. This type of feeding may be especially evident on apple trees and on raspberry and blackberry canes. Turkeys leave tell-tale inverted V-shaped scratchings in the leaves in areas where they forage for mast. Deer and squirrels also paw through leaves for mast. Partially eaten apples and other fruits are a distinctive sign of wildlife activity—the size of the tooth marks is a clue to the identity of the animal that was feeding on the fruit.

Scats

The fecal droppings, or *scats*, of various animals are sure clues to the identity of animals occupying an area. Scats are often found along trails and stream banks and may show remains of food items. Again, several useful field guides are available.

Other Signs of Wildlife Presence

The following list provides additional signs which are useful for identifying wildlife activity:

- Trails and runways
- Nests, dens, and burrows
- Claw marks on logs and the bark of trees
- Feathers
- Fur caught on trees and fences
- Dusting spots (grouse, flickers, and other birds)
- Remains of prey
- Holes dug in search of food
- Overturned rocks and ripped-up logs or stumps (skunks and black bears)
- Deer beds (look for matted vegetation, tracks, droppings, and hair)
- Debarked saplings ("buck rubs"—where male deer have rubbed their antlers)
- Scents, scent posts, and marking posts

The presence of certain species can be determined by signs of their feeding activities.

Table 3. Some seed characteristics of commercially important North American forest trees.

Most American forest trees reproduce primarily from seeds. To restore and maintain good forest stands, it is necessary to provide favorable conditions for natural seeding or to gather and sow seeds either directly on the land or in nurseries to produce seedlings which can be planted. This requires an understanding of seed habits and characteristics. Most tree seeds provide food for many birds and small mammals. Many of the berry, fleshy fruit, and nut species also are a source of food for larger mammals, including man.

Species	Time of Flowering	Time of Seed dispersal	Commercial seedbearing age[1] (Years)	Frequency of good seed crops[2] (Years)	Weight per 1,000 cleaned seeds[3] (Grams)	Average laboratory germination[4] (Days)
Ash, green (*Fraxinus pennsylvanica*)	May	Oct.–May	20–?	1+	26.22	42
Ash, white (*F. americana*)	Apr.–May	Sept.–Dec.	20–175	3–5	45.36	38
Aspen, quaking (*Populus tremuloides*)	Apr.–May	May–June	20–70+	4–5	.13	59
Basswood (*Tilia americana*)	June–July	Fall–spring	15–100+	1+	90.70	94
Beech, American (*Fagus grandifolia*)	Apr.–May	After first heavy frost	40–?	2–3	263.50	85
Birch, sweet (*Betula lenta*)	Apr.–May	Sept.–Nov.	40–?	1–2	.70	43
Birch, gray (*B. populifolia*)	Apr.–May	Oct. Jan.	8–50	1+	.11	64
Birch, yellow (*B. alleghaniensis*)	Apr.–May	Nov.–Feb.		1–2	1.01	27
Birch, paper (*B. papyrifera*)	Apr.–June	Sept.–Apr.	15–70+	1+	.33	34
Boxelder (*Acer negundo*)	Mar.–May	Sept.–Mar.		1+	38.44	33
Butternut (*Juglans cinerea*)	Apr.–May	Sept.–Oct.	20–80	2–3	15,120.00	65
Catalpa, northern (*Catalpa speciosa*)	May–June	Oct.–Mar.	20–?	2+	21.60	75
Cherry, black (*Prunus serotina*)	Mar.–June	June–Oct.	10–125	1+	94.50	63
Chestnut, American (*Castanea dentata*)	June–July	Oct. Nov.		1+	3,489.23	72
Cottonwood, eastern (*Populus deltoides*)	Feb.–May	Apr.–June	10–death	1+	1.30	88
Elm, American (*Ulmus americana*)	Feb.–Apr.	Mar.–June	15–300	1+	6.67	63
Elm, rock (*U. thomasii*)	Mar.–May	May–July	20–250	3–4	64.80	85
Elm, slippery (*U. rubra*)	Feb.–Apr.	Apr.–June	15–200	2–4	11.06	17
Fir, balsam (*Abies balsamea*)	May	Sept. Nov.	20–60+	2–4	7.56	22
Hackberry (*Celtis occidentalis*)	Apr.–May	Oct.–winter		1+	105.49	41
Hemlock, eastern (*Tsuga canadensis*)	May–June	Sept.–winter	30–400+	2–3	2.43	38
Hickory, bitternut (*Carya cordiformis*)	Apr.–May	Sept.–Dec.	30–175	3–5	2,907.69	55
Hickory, mockernut (*C. tomentosa*)	Apr.–May	Sept.–Dec.	25–200	2–3	5,040.00	66
Hickory, pignut (*C. glabra*)	Apr.–May	Sept.–Dec.	30–300	1–2	2,268.00	85
Hickory, shagbark (*C. ovata*)	Apr.–June	Sept.–Dec.	40–300	1–3	4,536.00	80
Honeylocust (*Gleditsia triacanthos*)	May–June	Sept.–Feb.	10–100	1–2	162.00	50
Locust, black (*Robinia pseudoacacia*)	May–June	Sept.–Apr.	6–60	1–2	18.90	68
Maple, red (*Acer rubrum*)	Feb.–May	Apr.–July		1+	19.89	46
Maple, silver (*A. saccharinum*)	Feb.–Apr.	Apr.–June	35–?	1+	324.00	76
Maple, sugar (*A. saccharum*)	Mar.–May	Oct.–Dec.	?–200+	3–7	74.36	39
Oak, black (*Quercus velutina*)	Apr.–May	Sept.–Nov.	20–100	2–3	1,814.40	47
Oak, bur (*Q. macrocarpa*)	Apr.–May	Aug.–Sept.	35–400	2–3	6,048.00	45
Oak, chestnut (*Q. prinus*)	Apr.–May	Sept.–Nov.	20–150	1–2	6,048.00	82
Oak, northern red (*Q. rubra*)	Apr.–May	Sept.–Oct.	25–200	2–3	3,240.00	58
Oak, scarlet (*Q. coccinea*)	Apr.–May	Sept.–Oct.	20–150	Irregular	1,620.00	62
Oak, white (*Q. alba*)	Apr.–May	Sept.–Oct.	20–300	4–10	3,024.00	78
Pine, eastern white (*Pinus strobus*)	Apr.–June	Sept.–Oct.	15–250	3–5	16.80	64
Pine, jack (*P. banksiana*)	May	Fall–several years	5–80+	3–4	3.45	68
Pine, red (*P. resinosa*)	Apr.–May	Fall–summer	25–200+	3–7	8.72	75
Pine, shortleaf (*P. echinata*)	Mar.–Apr.	Nov.–Dec.	16–280+	5–10	9.45	68
Red cedar, eastern (*Juniperus virginiana*)	Mar.–May	Feb.–Mar.	10–175	2–3	10.50	42
Spruce, black (*Picea mariana*)	May–June	Oct.[5]	30–250	4–5	1.12	64
Spruce, red (*P. rubens*)	Apr.–May	Sept.	30–?	3–8	3.24	60
Spruce, white (*P. glauca*)	May	Aug.–Nov.	30–?	2–6	1.89	50
Sweetgum (*Liquidambar styraciflua*)	Mar.–May	Sept.–Nov.	20–150	1–3	5.53	70
Sycamore, American (*Platanus occidentalis*)	May	Sept.–May	25–250	1–2	2.22	35
Tamarack (*Larix laricina*)	May	Sept.	40–75+	5–6	1.42	47
Tupelo, black (*Nyssa sylvatica*)	Apr.–June	Sept.–Oct.			137.45	30
Walnut, black (*Juglans nigra*)	May–June	Fall	12–?	Irregular	11,340.00	75
White-cedar, Atlantic (*Chamaecyparis thyoides*)	Mar.–Apr.	Oct.–Nov.	4–100+	1+	.99	84
White-cedar, northern (*Thuja occidentalis*)	Apr.–May	Aug.–Oct.	30–100+	5	1.31	46
Yellow-poplar (*Liriodendron tulipifera*)	Apr.–June	Oct.–Jan.	15–200+	Irregular	32.40	5

1. Most tree species begin to bear seeds several years earlier than indicated and continue almost to death, but the most abundant production normally is between the ages indicated. Open-grown trees usually bear earlier and more abundantly than those in stands.
2. Most trees bear some seed in the years between good crops, although total failures may occur.
3. Seeds cleaned for commercial use. Wings and fleshy parts removed in many species.
4. Seeds of many woody plants contain dormant embryos. Such dormancy usually can be broken by holding the seeds for one to three months in a moist medium at 0°–5°C. Some species, chiefly legumes, have hard or impermeable seedcoats which can be overcome by mechanical scarification or soaking in H_2O. Several trees have seeds with both types of dormancy. In some species, seed dormancy is general; others may have both dormant and nondormant seeds in the same lot; and still others may vary between lots, some lots being completely dormant and others completely nondormant.
5. Black spruce cones are retained for two to three years in a state of active seed dispersal.

Adapted from the 1961 USDA Yearbook of Agriculture

Chapter 2: Understanding Wildlife Habitats

Table 4. Status, relative abundance, and home-range requirements of selected New England mammalian species.[a]

Name	Population status in New England[b]	Density (average no. per area)	Average home-range requirement
Short-tailed shrew (*Blarina brevicauda*)	C	Maximum 48 per acre	0.5–1.25 acres
Eastern mole (*Scalopus aquaticus*)	LC	1 per 3–5 acres	0.3 acre
Big brown bat (*Eptesicus fuscus*)	C	12–200 per colony	Probably <30 miles from birthplace
Opossum (*Didelphis marsupialis*)	C–U	9.1 per kilometer^2 (2–125 per kilometer^2)	Minimum 11.5 acres
New England cottontail (*Sylvilagus transitionalis*)	U–R	(Data not available)	0.5–1.8 acres
Snowshoe hare (*Lepus americanus*)	C in suitable habitat	Approximately 100 per km^2	Probably 10 acres
Woodchuck (*Marmota monax*)	C	2–16 per hectare	0.25–0.5 mile in diam.
Gray squirrel (*Sciurus carolinensis*)	C–A	(Data not available)	2–7 acres
Red squirrel (*Tamiasciurus hudsonicus*)	C–U	0.3–1.5 per hectare	2.73–6.03 acres < 200 yards in diam.
Southern flying squirrel (*Glaucomys volans*)	C–U	Maximum 5 per acre	Males: 0.41 acres females: 0.53 acres
Eastern chipmunk (*Tamias striatus*)	C–U	2–4 per acre (Max: 30 per acre)	0.5–1.0 acre
Beaver (*Castor canandensis*)	C	0.5 per acre (5–8 per colony)	450 feet from water
White-footed mouse (*Peromyscus leucopus*)	C	Maximum 15 per acre	Males: 0.16–0.54 acre females: 0.06–0.36 acre
Boreal red-backed vole (*Clethrionomys gapperi*)	C	Maximum 10 per acre	0.25 acre
Meadow vole (*Microtus pennsylvanicus*)	A	35 per acre (Max: 50–100 per acre)	0.08–0.23 acre
Muskrat (*Ondatra zibethicus*)	C–U	10 per acre (fall)	200 yards of den
Porcupine (*Erethizon dorsatum*)	C–U	(Data not available)	Summer: 35 acres winter: 10 acres
Coyote (*Canis latrans*)	U–C	1 per 2 miles^2	Females: 10–30 miles^2 males: 40 miles^2
Red fox (*Vulpes fulva*)	C–U	0.01 per acre	1–2 miles^2
Gray fox (*Urocyon cinereoargenteus*)	C–U	0.01 per acre	1–5 miles^2
Black bear (*Ursus americanus*)	C (north) U (south)	Maximum 1 per 1.3 miles^2	Females: 1–10 miles^2 males: 30–40 miles^2
Raccoon (*Procyon lotor*)	C	1 per 4.4–47 acres	0.6–1.8 miles^2
Long-tailed weasel (*Mustela frenata*)	C–U	1 per 15.5–18.1 miles^2	29.6–39.5 acres
Mink (*Mustela vison*)	C–U	(Data not available)	2–3 miles^2
Striped skunk (*Mephitis mephitis*)	C	31 per mile^2	0.25–0.50 mile^2
River otter (*Lutra canadensis*)	U	(Data not available)	15 or more linear miles
Bobcat (*Felis rufus*)	C–U	1 per 2–4 miles^2	0.4 to 15.8 miles^2
White-tailed deer (*Odocoileus virginianus*)	C	Maximum 100 per mile^2	150–500 acres
Moose (*Alces alces*)	LC–U	1/5 miles^2 (2 per mile^2 approaching K)	2–10 miles^2

a. Compiled by J. S. Barclay and C. Giambartolomei-Green. References:
  DeGraaf, R. M. and D. D. Rudis. 1986. *New England Wildlife: Habitat, Natural History, and Distribution.*
  Burt, R. M. and R. P. Gossenheider. 1976. *A Field Guide to the Mammals, North American and North of Mexico.*
  Timm, R. M., ed. 1983. *Prevention and Control of Wildlife Damage.*
  Schmidt, J. L. and D. L. Gilbert, eds. 1978. *Big Game of North America.*

b. A = abundant; C = common; LC = locally common; U = uncommon; R = rare

Review Questions

1. What is wildlife habitat? What are the three primary elements of wildlife habitat?

2. What is a limiting factor?

3. Give three examples of important wildlife cover.

4. What is mast? Give five examples of trees or shrubs that produce mast.

5. What is territoriality? Why do some wildlife species exhibit territoriality?

Chapter 2: Understanding Wildlife Habitats

Field Exercises

1. Locate three different examples of wildlife cover on your woodlot.

    Are there ways that you could improve or expand each of them?

2. Using table 3 on page 33, locate a good mast-producing area on your woodlot.

    What condition do the mast-producing plants appear to be in, i.e., are they healthy and will they remain so for several years, or are they crowded and overtopped by other competing species?

    Make a list of ways that you could improve their health or encourage their growth.

3. With a photocopy of the forest cover type map that you made for chapter 1, walk through your woodlot and sketch the following elements of wildlife habitat.

    a. Sketch any streams, ponds, spring seeps, or other water sources.

    b. Sketch important wildlife cover areas, paying particular attention to dense brush or conifers, ledges, caves, stonewalls, and areas containing numerous cavity trees or snags.

    c. Mark significant mast and other food-producing areas. Also, pinpoint apple trees and wolf trees if they are present.

    d. Designate edges, e.g., between fields and woods or between forested areas of noticeably different ages.

    Examine your map.

    Are there areas where cover, food, and water are all available within a relatively small area?

    Do you see areas where two of these three elements are available?

    What seems to be the limiting factor in areas with fewer than three elements present?

    Retain a master copy of this map and make several photocopies for use in the remaining chapters. The photocopies will form the basis for future field exercises as you develop your wildlife management plan.

4. Walk through a portion of your woodlot and look for wildlife signs, e.g., tracks, scats, trails, runways, and evidence of feeding.

    Do you notice that more signs are present in some areas than in others?

    If so, do these areas contain cover, food, and water in close proximity; or do they contain only one frequently visited habitat element?

These exercises were designed to be done on your own woodland. Once completed, you will be well on your way toward your own wildlife management plan.

5. Make a track bed for detecting tracks by spreading and moistening sand or by raking soft ground at areas where wildlife sign is prevalent. Although not necessary, the bed can be baited with different foods to further attract species. For example, lettuce and carrots can be put out one day, peanut butter on crackers the next day, and sardines the following day. Try establishing several beds in areas of different stages of forest succession and along streams and ponds.

   Do the different baits attract different species, i.e., herbivores vs. carnivores or omnivores?

   Do certain species tend towards areas in particular stages of forest succession?

Field Notes:

3 American Woodcock and Ruffed Grouse

Northeastern forests have matured, the acreage of essential woodcock and grouse habitats in early and intermediate successional stages has greatly diminished, and the numbers of these popular upland gamebirds have declined substantially.

Introduction

By the end of World War I, forests in the northeastern United States had been cut heavily. The early successional forests that resulted provided an abundance of essential habitats for American woodcock and ruffed grouse. Now, some seventy years later, northeastern forests have matured, the acreage of essential woodcock and grouse habitats in early and intermediate successional stages has greatly diminished, and the numbers of these popular upland gamebirds have declined substantially.

This chapter discusses management practices designed to enhance the ability of forested sites to support breeding populations of American woodcock and ruffed grouse. A full chapter is devoted to these two species because some of the habitat management practices implemented for them stratify the forest into several age classes. Small portions of forestland are returned to early and intermediate stages of forest succession, providing habitats for the numerous wildlife species that prefer these to uniform tracts of mature forest.

American Woodcock

Description and Range

The American woodcock *(Scolopax minor)* is a member of the sandpiper family (family Scolopacidae). Some other members of this family include the common snipe, godwits, curlews, dowitchers, and sandpipers. The woodcock's scientific name until recently was *Philohela minor*, meaning "little lover of swamps or bogs." Locally the bird may be known as timberdoodle, bogsucker, or Labrador twister.

Although most sandpipers inhabit marine and inland shoreline habitats, woodcock have adapted to forested habitats. They have a mottled brown appearance which blends with the dry leaf pattern of the forest floor, and their short, rounded wings enable them to fly in dense, brushy cover. Woodcock have short legs, large eyes set high and far back on the head, dark bands on top of the head, and a slender bill approximately 2.5 inches long (Owen et al. 1977) (figure 21).

American woodcock range from southern Canada to the Gulf states and from the Atlantic coast westward into Texas, Oklahoma, and the eastern edge of the central states and provinces (Smith and Barclay 1978) (figure 22). Most woodcock breed in the northern half of their range and have migrated, by winter, to the southeastern United States.

The American woodcock is a member of the sandpiper family.... Although most sandpipers inhabit marine and inland shoreline habitats, woodcock have adapted to forested habitats.

Figure 21. The woodcock is the only member of the sandpiper family adapted to forest rather than shoreline habitats.

Illustration: Jim Lish

Life History

Spring Migration

Woodcock generally leave the southern part of their wintering range in late January or early February and arrive in New England from mid-March to mid-April to breed (Sepik et al. 1981). Males tend to migrate first, and early arrivals of both sexes frequently encounter late snowstorms or unmelted snow (Owen et al. 1977).

Courtship

While in transit and upon arrival at their northerly breeding ranges, male woodcock perform a series of courtship displays on areas called **singing grounds**, which typically are small herbaceous clearings in brushy fields.

Courtship rituals consist of aerial and ground displays performed at dawn and dusk. The aerial display is a flight above the singing ground lasting forty to sixty seconds (Owen et al. 1977). The male flies upward in a large spiral, making a distinct twittering sound produced by the three outer **primary feathers** of each wing (figure 23). After spiraling higher in tightening circles and flying almost out of sight, the male reaches his highest elevation and plunges rapidly downward in a number of steep, "J"-shaped dives, producing a

Courtship rituals consist of aerial and ground displays performed at dawn and dusk.

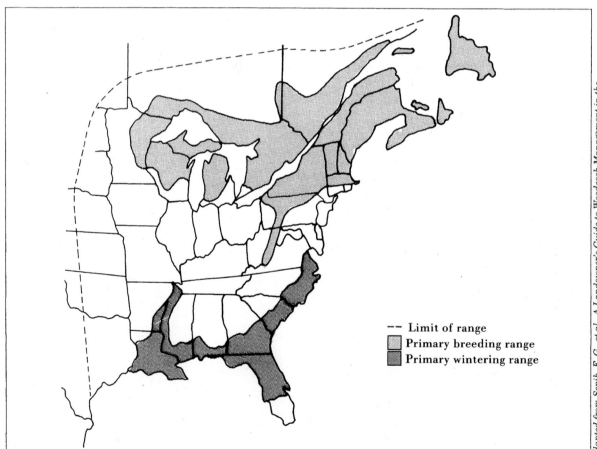

Figure 22. Woodcock have been known to breed as far north as central Canada and to winter as far south as central Florida.

Adapted from Sepik, F. G., et al., *A Landowner's Guide to Woodcock Management in the Northeast*; and Smith, R. W., and J. S. Barclay 1978, "Evidence of Westward Changes in the Range of the American Woodcock" in *American Birds* 32(6): 1122–1127.

series of musical chirps. Upon landing, he performs the ground display which includes occasional strutting and regular vocalizations resembling a nasal, buzzy "peent" sound.

Each session of displays usually lasts from thirty to sixty minutes and includes ten to twenty aerial flights (Owen et al. 1977). Besides attracting females, the displays advertise the presence of a male to other males. Older males are territorial and ward off younger males with display threats and aggressive aerial pursuits accompanied by loud cackling.

Nesting

Woodcock typically begin nesting in April throughout New England. The nest site is located on the ground, often within 400 feet of the singing site where courtship occurred, and is commonly placed within a few inches or feet of a sapling or similar **guard object** (figure 24). The nest is a shallow depression in the leaf litter, difficult to locate and usually found accidentally. Incubating females rely upon the protective coloration of their feathers for concealment from predators.

The four, slightly glossy eggs hatch after an incubation period of nineteen to twenty-two days. The eggs range in color from pinkish buff to cinnamon and are covered with brown, blotchy markings. Nests with two or three eggs are sometimes found late in the nesting season and may indicate renesting attempts after earlier nests have been disturbed.

Each session of displays usually lasts from thirty to sixty minutes and includes ten to twenty aerial flights.

Figure 23. Woodcock are best known for their unique spring courtship displays. The male woodcock's primary feathers produce a distinct twittering sound as they fly upward in ever tightening spirals.

Chapter 3: American Woodcock and Ruffed Grouse

Brood Rearing

Apparently, the female incubates the eggs and rears the young with no noticeable assistance from the male (Vermont Fish and Wildlife 1986). Young woodcock are **precocial**, meaning that they hatch fully feathered with eyes open and legs capable of running rapidly. Chicks leave the nest soon after hatching, fly short distances by two weeks, and are nearly full grown at one month. By early June, most young become completely independent, but stay in the vicinity of the nesting site until fall migration (Sepik et al. 1981).

Fall Migration

Fall migration begins with the onset of the first heavy frosts in October and continues through November. During mild autumns, **flights** will drift casually southward stopping to rest and feed, but during cold, severe autumns, woodcock often migrate rapidly and continuously.

The arrival of flights at temporary resting and feeding sites may be determined by periodically visiting unfrozen feeding areas, such as spring seeps and poorly drained (soggy), brushy lowlands. If a number of woodcock are observed one day when none were observed previously, then a flight has probably arrived (Sheldon 1971).

Feeding

Most of a woodcock's feeding time is spent probing in moist soils. The upper mandible of the bill is slightly flexible at the tip, allowing

> *Chicks leave the nest soon after hatching, fly short distances by two weeks, and are nearly full grown at one month.*

Figure 24. The woodcock's nest is simply a shallow depression in the leaf litter, located close to a sapling or other guard object.

the woodcock to grasp underground food (figure 25). Fifty to ninety percent of the diet consists of earthworms (Sepik et al. 1981), but beetle larvae and fly larvae are consumed regularly. Woodcock diets occasionally include ants, moths, snails, and seeds from various plants (Sheldon 1971).

Roosting

Woodcock roost on the ground, and many spend summer nights resting in **roosting fields**. Singing grounds and abandoned agricultural fields near daytime covers are common roosting sites. Woodcock fly to the fields at dusk and return to nearby feeding covers at dawn (Sepik et al. 1981). In fields used by many woodcock, small groups will often congregate in shallow depressions—perhaps to assist in detecting predators (Connors and Doerr 1982) or for warmth.

Longevity

The average longevity (life span) for woodcock is 10.6 months (Blankenship 1957), although many woodcock live for several years and some have been recaptured in their seventh year.

External Sex Characteristics

Female woodcock typically are larger and heavier than males, and both sexes are proportionally heavier in the fall than in the spring—probably due to increased fat reserves for migration. Individuals with bills 2.72 inches (69 millimeters) or longer are usually females, and those with bills 2.64 inches (67 millimeters) or shorter are usually males. The combined widths of the three outer primary feathers of the wing (measured 2 centimeters from the tip of each feather) are narrower (less than 12.4 millimeters) in the male than in the female (greater than 12.6 millimeters).

Fifty to ninety percent of the diet consists of earthworms, but beetle larvae and fly larvae are consumed regularly.

Figure 25. The woodcock's bill is adapted for probing the soil and grasping underground food.

Habitat Requirements

Woodcock prefer the particular conditions associated with early stages of forest succession. Abandoned agricultural fields, pastures, brushy lowlands, and wet areas provide essential habitats for courtship, nesting, feeding, and roosting. As these areas mature, they become less attractive to woodcock.

Courtship Habitat

Singing grounds are the critical habitat component for courtship—they provide display sites for males and mating opportunities for the pair. Clearings used as display sites are typically located in brushy fields of 1–3 acres (figure 26). Sweet fern, blueberry, and young stems of gray birch, aspen, black cherry, white pine, red cedar, juniper, alder, red maple, and dogwood are commonly found on sites suitable as singing grounds in the Northeast. However, as woody vegetation becomes increasingly dense and the clearings used for displaying become overgrown (as forest succession proceeds) the suitability of the site for singing grounds diminishes. Hiding places develop for ground predators such as foxes and cats, and display flights become hampered (Gutzwiller and Wakeley 1981). Fields lacking woody vegetation and clumps of tall, herbaceous plants are often avoided; scattered vegetation apparently provides some protective screening from predators such as hawks and owls.

Proximity of the singing ground to moist soils is important. Singing sites which appear suitable otherwise, may be vacant if they are not located near feeding sites where soils are easily probed for food

As woody vegetation becomes increasingly dense and the clearings used for displaying become overgrown (as forest succession proceeds) the suitability of the site for singing grounds diminishes.

Figure 26. Singing grounds are a critical habitat component. Clearings in brushy fields close to moist soils make the best sites.

(Lambert and Barclay 1976). The height of surrounding trees is also important. Small fields surrounded by tall trees may not be used if the trees interfere with the initial spiral of the display flight (Sheldon 1971).

When suitable singing grounds are in short supply, or when surplus males are present, unusual sites may be used for courtship displays. Woodcock have been observed displaying on cultivated fields, lawns, and baseball diamonds (Liscinsky 1972).

Nesting Habitat

Woodcock nest in a variety of cover types, generally less than 400 feet from occupied singing grounds. Preferred nesting sites are in young, open woodlands with a scattered understory of brush and seedlings (figure 27). Nests may also be found in fields containing singing grounds or along the edges of mature forests, depending upon the size of the fields and the age of surrounding cover types. Higher dry areas are often favored for nesting over poorly drained lowland areas (Liscinsky 1972).

Feeding and Brood-rearing Habitat

Good feeding sites are perhaps the most critical habitat component for attracting and managing woodcock. Although the appearance of feeding sites can vary considerably, favorable feeding sites are often moist (not wet), lowland sites with moderately dense brush and few **overtopping trees**. In New England, alder (*Alnus rugosa*) stands can be one of the best indicators of an appropriate

Although the appearance of feeding sites can vary considerably, favorable feeding sites are often moist (not wet), lowland sites with moderately dense brush and few overtopping trees.

Figure 27. Preferred nesting sites are in young, open woodlands with a scattered understory of brush and seedlings.

woodcock feeding habitat (figure 28). Alders characteristically grow in and around poorly drained sites, and they have nodules on their roots which host a specific bacteria capable of fixing nitrogen in the soil. This combination of moister sites and nitrogen-rich soils often favors large populations of earthworms. The presence of alders does not guarantee the presence of woodcock, however. Soils may be either too acid or too alkaline, contain too much sand or silt, or hold too much or too little moisture to support earthworms and woodcock (Liscinsky 1972).

Woodcock feed in various woodland habitats that lack alder. Stands of young aspen, red maple, and some dogwoods and willows can be highly indicative of potential feeding habitats. If a healthy population of earthworms is absent, however, then woodcock will usually be absent as well.

A brood-rearing habitat is essentially the same as a feeding habitat except that areas with bare ground or dense ground cover are initially avoided (Sepik et al. 1981), partly due to the vulnerability of the chicks to predators. Females lead their broods to moist areas where chicks can probe easily for food (Sheldon 1971).

Woodcock roost on the ground, so the presence of appropriate nighttime cover is important.

Roosting Habitat

Woodcock roost on the ground, so the presence of appropriate nighttime cover is important. Fields or clearings of at least 3 acres and with low-growing vegetation are preferred for roosting. Pastures, abandoned fields, hay fields on poor soils, low-bush blueberry fields, and recently harvested woodlots are commonly used (Sepik et al. 1981). Lush hay fields and actively farmed fields that have been plowed, disked, or planted in cover crops are rarely used (Connors and Doerr 1982). Abandoned agricultural fields with eastern red cedar,

Figure 28. Dense alder stands make excellent woodcock feeding sites.

juniper, meadowsweet, sweet fern, and other similar species are often ideal. Woodcock usually roost within ½ mile of daytime feeding sites.

Are Woodcock Using Your Woodlot?

One of the easiest ways to detect woodcock presence in the spring is to listen during the evening after sunset for the nasal "peent" or "bzeep" and then for the twittering sound as the male executes his display flight. Twittering sounds can also be heard when birds are flushed or as they fly low from daytime feeding and resting cover to singing grounds. Evidence of feeding can be detected by the presence of probe holes and "splashes" (figure 29). Probe holes are small oval or dumbbell-shaped holes in soft soil made when woodcock insert their bills to obtain food. Splashes are the bird's droppings and appear as large spots of thinned white paint amongst probe holes. Starlings and snipe leave similar evidence, but their feeding locations generally do not overlap with those of woodcock. Soils surrounding spring seeps are likely sites to search for probe holes and splashes.

Managing Habitats for Woodcock

The following sections on managing habitats for woodcock are based primarily upon the 1981 publication *A Landowner's Guide to Woodcock Management in the Northeast*, by G. F. Sepik, R. B. Owen, and M. W. Coulter. This publication was compiled after extensive woodcock research at the Moosehorn National Wildlife Refuge in Calais, Maine. Although the research was conducted in habitats largely comprised of alder, spruce, and fir, the concepts and management strategies are applicable to forests throughout the Northeast that may lack a prevalent alder component.

One of the easiest ways to detect woodcock presence in the spring is to listen during the evening after sunset for the nasal "peent" or "bzeep" and then for the twittering sound as the male executes his display flight.

Figure 29. Probe holes are clear evidence that woodcock have been feeding in an area.

When managing for a particular species of wildlife, care must be exercised to determine if a site is suitable for that species. Not all areas lend well to conversion into woodcock habitat, such as property located on dry ridgetops. Working with a habitat that is, or was, suitable for woodcock is most desirable, but do not be discouraged if such a habitat is not immediately apparent. By walking through a site and looking for some of the elements required by woodcock, one can assess its potential for successful woodcock management.

Singing Ground Management

Woodcock use of an area can often be dramatically increased by creating singing grounds where few are present. One singing ground for every 5–10 acres of suitable land is generally adequate for encouraging high local breeding densities while minimizing competition for display sites. Appropriate sites should be carefully chosen. Relatively flat areas near, or within, feeding covers are excellent places to establish singing grounds. Provided that feeding cover is nearby, many woodland sites can be successfully modified into singing grounds regardless of the successional stage and tree species present. Old, brushy fields are especially desirable to manage as singing grounds.

Singing grounds in wooded areas can be created by patch cutting (***clear-cutting***), i.e., the removal of all trees within a small forest area. Where surrounding trees are taller than 25 feet, patch cuts should be at least 0.5 acres, but where trees are shorter, patch cuts can be as small as 0.25 acres. The removal of logs and tops is desirable, but not essential. If felled trees are to be left on the site, the directions of fall can be manipulated so that several areas of open ground remain for displaying males. Young clumps of alder, aspen, and dogwood should not be cut.

Woodcock use of an area can often be dramatically increased by creating singing grounds where few are present.

Figure 30. Woodcock are most attracted to singing grounds of at least 500 square feet, managed to limit the height of trees and shrubs around the perimeter.

Singing grounds in existing fields can be created by mowing, disking, **brush hogging**, chain sawing, or bulldozing—depending upon the amount and age of woody vegetation. Cleared sites should be at least 100 feet in diameter, and no further than 600 feet from tree cover, borders, streams, or other substantial edges. Singing grounds should be no closer to each other than 1,000 feet in large fields where several sites can be cleared (Lambert and Barclay 1976).

The shape of clearings is not critical to woodcock once size requirements have been satisfied, but a greater number of wildlife species will benefit if the principles of edge are considered. Patch cuts shaped in irregular ovals or rectangles have a greater amount of edge, and thus edge habitat, than circles or squares of equal area.

An ideal singing ground plan might consist of keeping at least 500 square feet free of brush and trees. Keep brush no taller than 2 feet for at least 50 feet around the clearing and trees no taller than 30 feet for an additional 50 feet (figure 30).

The useful life of singing grounds vary. Clearings that are cut in hardwood stands may sprout prolifically and rapidly reduce the attractiveness to woodcock. Sprout growth can be slowed by cutting trees in the summer or controlled by the use of herbicides, prescribed fire, or by periodic cutting. (Restrictions on herbicides and prescribed fire vary locally, so regulations on their use should always be consulted.) If singing grounds are created periodically, there is less need for annual maintenance, and the reverting, older sites may become desirable feeding, nesting, and brood-rearing cover.

Many other wildlife species benefit from the establishment of carefully situated singing grounds. Young ruffed grouse and wild turkey depend upon woodland clearings for supplying a summer diet of insects. Many songbirds require such habitats for breeding, and deer browse heavily on the incoming sprouts. Also, firewood and timber can be obtained from woodland patch cuts (Lyons and Broderick 1986).

Feeding, Nesting, and Brood-rearing Cover Management

Sites managed as feeding (daytime) covers should be located on soils capable of supporting healthy earthworm populations. Soils such as muck soils or dry, sandy soils usually support few earthworms and, therefore, have low potential. Soil surveys (published by the USDA Soil Conservation Service) can be helpful for identifying suitable soil types. Working with existing alder stands can be the most productive approach to feeding cover management. Although alders grow rapidly and require periodic maintenance, they are relatively sensitive to moisture differences in the soil and are usually good indicators of favorable sites to manage. If alder is not present on a particular site, other woody species can be effectively managed to serve well as feeding cover.

On sites dominated by alders or other hardwoods less than twenty years old, strips approximately 70 feet wide can be clear-cut and spaced approximately 280 feet apart. New strips 70 feet wide should be cleared next to old strips every four years so that each strip is cut and replaced at twenty-year intervals, or **rotations**. The length, shape, and number of strips can be varied to accommodate small areas as long as twenty-year rotations are maintained. Alder and other

Sites managed as feeding (daytime) covers should be located on soils capable of supporting healthy earthworm populations.

hardwoods **regenerate** in the clearcut strips, and the overall area will contain patches of varying ages (figure 31). If obtaining firewood is a major objective, however, the rotation schedule can be adjusted to twenty-five years—an ideal rotation length for cordwood production on high quality sites (Lyons and Broderick 1986). Allowing hardwoods to reach merchantable size can also help defray the cost of clearing.

Maturing alders and invading trees eventually decrease the attractiveness of the clearcuts for feeding, but as the number of stems decreases (when managed strips approach their next cutting cycle), suitability for nesting and initial brood rearing increases (figure 32). When clearing strips to regenerate feeding sites, cutting should be accomplished during the dormant season—after leaf fall but before bud break (in autumn and winter). Otherwise, the root systems of the cut trees and shrubs might become stressed and either die or produce fewer or weaker sprouts.

Alders serve as optimal feeding sites for approximately ten years, and by thirty years they are generally overtopped by other trees. On sites dominated by alders greater than twenty years old (3–5 inches in diameter), 70-foot strips should be cleared on a schedule such that the entire area is cut in ten years—unlike the twenty-year rotation for young alder. After ten years, the overall area will be composed of young alder, and the strips can then be cut and replaced on the twenty-year rotation described above. In areas where alder is overtopped by trees and in danger of dying out, the entire site should be cleared before a rotation schedule can be followed.

Alders serve as optimal feeding sites for approximately ten years, and by thirty years they are generally overtopped by other trees.

Figure 31. Feeding sites are best managed on twenty- to twenty-five-year rotations, to maintain brushy patches of varying ages. New stripcut (left), stripcut one year later (right).

If the management site is sloped, the strips should follow the contours to minimize soil erosion. Spring seeps, other wet areas, and small *intermittent streams* can be included in the clearcuts.* Alders rejuvenate rapidly in low, moist places, thus providing varying heights and densities of alder within the strip while creating feeding sites late in the summer when the surface of higher ground becomes too dry to support earthworms. Streams with year-round flows should be protected by an uncut *buffer strip* on each side at least three times the stream's width to avoid unnecessary *siltation* and increased summer temperatures. Buffer strips can be rejuvenated by periodically *thinning* the oldest stems.

Young aspen and red maple stands located on rich soils can also serve as good feeding sites. Clear-cutting after leaf fall encourages aspen and red maple to sprout prolifically from their roots, often resulting in highly favorable feeding sites. As aspen and red maple stands grow older (approaching fifteen years) they provide excellent nesting and initial brood-rearing habitat and also are attractive to ruffed grouse and deer.

Stands of young conifers near feeding habitats can provide important daytime cover during warm weather and extended dry periods. However, an effort may be necessary to prevent them from invading and overtopping the valuable feeding sites.

*A*lders rejuvenate rapidly in low, moist sections, thus providing varying heights and densities of alder within the strip while creating feeding sites late in the summer. . . .

---

* Some jurisdictions require a permit for the clear-cutting of timber in wetlands.

Figure 32. As feeding sites approach maturity, their suitability for nesting and brood rearing increases.

Figure 33. Pastures and lowbush blueberry fields make excellent roosting areas.

Managing Roosting Sites

Traditional agricultural practices will suffice to provide pastures, hayfields, and blueberry fields used for roosting (figure 33). However, abandoned agricultural land will revert to woodland by the process of old field succession and become unsuitable for roosting habitat. If at least one 3-acre roosting field is present per 100 acres of reverting farmland or uncleared land, then abandoned fields and pastures can be managed for feeding, nesting, and brood rearing. Otherwise, they should be partially cleared for roosting by the methods previously described in the "Singing Ground Management" section.

If fields of adequate size are not available, then small fields can be enlarged or new ones can be created. Where there is a choice, it is preferable to enlarge existing fields. Newly cleared fields may not be used by woodcock for several years and can be expensive and difficult to create, but enlarged fields will usually receive immediate use. It is desirable to make initial clearings larger than the intended field. In this manner, a strip 50–100 feet wide can be allowed to grow back around the field for future nesting and brood rearing.

Roosting fields can be maintained by herbicide application, prescribed burning, mowing, or cutting. Again, local regulations should be consulted before attempting to burn or use herbicides. Periodic mowing is effective for reducing invading woody vegetation and dense herbaceous growth in hayfields, pastures, and enlarged fields. When possible, mowing should be done in late summer, fall, or winter to avoid disturbance of ground-nesting birds and to provide seasonal singing sites.

Roosting fields can be maintained by herbicide application, prescribed burning, mowing, or cutting. Again, local regulations should be consulted before attempting to burn or use herbicides.

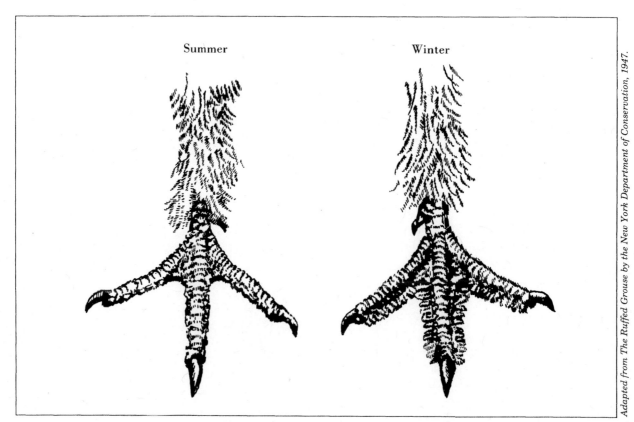

Figure 34. Ruffed grouse grow a temporary fringe on their toes for "snowshoeing" during the winter.

Ruffed Grouse

Description and Range

Ruffed grouse (*Bonasa umbellus*) belong to the family Tetraonidae which includes nine other species of grouse native to North America. In New England, ruffed grouse are known locally as partridge or patridge, but they are not closely related to true partridges.

Ruffed grouse are the size of small chickens and possess a large fanlike tail and two patches of iridescent black or dark brown neck feathers called a ruff. Two distinct color phases exist depending upon geographic location. Variations of gray predominate in colder northern and western regions of the range, and reddish brown predominates in southern regions. These colors overlap geographically, and both gray and red phase grouse inhabit the northeastern United States. Ruffed grouse grow a comb-like fringe on the sides of their toes each winter, thus providing temporary "snowshoes" that are shed the following spring (figure 34).

Ruffed grouse are nonmigratory and spend their entire lives in forested areas. They are present in all of the Canadian Provinces and in thirty-eight of our more northern and eastern states—occupying, with wild turkeys, a range larger than that of other nonmigratory gamebirds native to North America (figure 35).

Ruffed grouse are nonmigratory and spend their entire lives in forested areas.

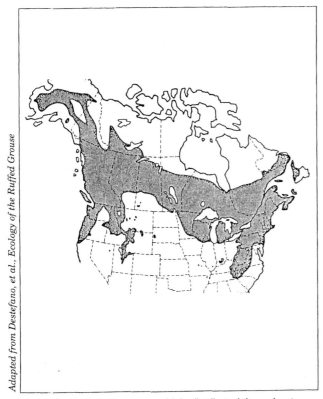

Figure 35. Ruffed grouse are widely distributed throughout much of Canada and the northern United States.

Figure 36. The "drumming" of the male ruffed grouse creates a distinctive and memorable sound.

53

Life History

Drumming and Courtship

Male ruffed grouse exhibit a behavior known as "drumming" to advertise their presence for courtship and territorial purposes. The male selects an elevated **drumming log** which may be a log, boulder, stonewall, or mound of soil on which he stands and rapidly flaps his wings (figure 36). Each wingbeat creates a vacuum that produces a muffled thumping sound audible through heavy cover. The thumping starts slowly and increases rapidly, lasting for approximately eight seconds and resembling the sound of a farm tractor being started in the distance.

Drumming occurs most frequently and vigorously in late March as the breeding season begins. Female grouse are attracted to the drumming site, and when a male sees a receptive female, he struts towards her with fanned tail, raised ruff, and lowered wings (figure 37). Upon approaching her, he stops, pivots his head from side to side with increasing speed, and then rushes towards her. Mating is brief, and they do not remain together. He generally remains at the drumming log while she departs and searches for a nesting site.

Males drum throughout the year as a means of claiming a territory. The territory is generally defended against the intrusion of other males and may encompass 6–10 acres of suitable habitat.

Drumming occurs most frequently and vigorously in late March as the breeding season begins.

Figure 37. The raised ruff of the courting male helped give the ruffed grouse its name.

Nesting

In the Northeast, ruffed grouse begin nesting in April. The nest site is commonly located at the base of a tree, stump, boulder, or ledge, and occasionally under brushpiles and logs. It is formed by the female and appears as a cup-shaped depression in the leaf litter. As is the case with woodcock, incubating females rely upon the dead leaf pattern of their feathers to conceal themselves and the nest from predators.

Females generally lay two eggs every three days until a clutch of eight to fourteen eggs results. To ensure that the chicks hatch at about the same time, the female does not begin incubation until all eggs have been laid. The eggs are a uniform pinkish-buff color occasionally spotted with dull brown. They hatch approximately twenty-five days after the last egg has been laid.

Brood Rearing

If newly hatched chicks are threatened by predators, the female often leaves the nest, pretending to have a broken wing. The predator follows her and, after dragging her "broken" wing for some distance, she flies away unharmed—presumably diverting the predator's attention from the chicks.

The chicks are precocial and leave the nest soon after hatching (figure 38). They begin to fly by one week, become nearly indistinguishable from the female by the end of summer and are fully independent by the end of September (Bump et al. 1947).

To ensure that the chicks hatch at about the same time, the female does not begin incubation until all eggs have been laid.

Figure 38. Chicks leave their nest shortly after hatching and begin to fly within a week.

Dispersal

Dispersal usually begins in mid-September and peaks in October (Chambers and Sharp 1958) as young grouse leave their family groups and search nearby for home ranges of their own. Young males often establish new territories less than 2 miles from the area in which they were raised. Once the male finds a suitable drumming log, he usually remains within 300 yards of the log for the rest of his life. Young females generally disperse later than the males and travel further (Gullion 1981).

As young grouse disperse, some exhibit what is known as the "crazy flight." They commonly fly into windows and other large, reflective objects as they travel through residential areas and other poor habitat situations.

Feeding

The diets of ruffed grouse vary noticeably in accordance with the seasons and geographic location. In the warmer months, greens, insects, berries, and other fruits are readily consumed. Through the cooler months, the diet shifts to **persistent fruits** and to buds of trees and shrubs. In regions supporting an abundance of aspen, the buds and **catkins** of mature male aspens are the most crucial cold weather food source (Svoboda and Gullion 1972). In locations where aspens are not prevalent, cold weather food sources such as buds and seeds of oaks, birches, beech, hawthorn, hornbeam, hop-hornbeam, shadbush,

> *Young males often establish new territories less than 2 miles from the area in which they were raised.... Young females generally disperse later than the males and travel further.*

Figure 39. Grapes, rose hips, and other "soft mast" are important seasonal grouse foods.

witch hazel, and hazel are heavily utilized. Examples of additional plants which are seasonally important grouse foods include dogwoods, apples, grapes, cherries, blackberries, strawberries, blueberries, barberries, winterberry, sumacs, viburnums, red maples, mountain laurel, rhododendrons, and roses (McDowell 1975; Bump et al. 1947) (figure 39).

Young grouse initially require a diet of insects, spiders, and other sources of animal protein for rapid development. As the young mature, they increasingly feed on plants, and by August their diet becomes similar to that of adults (Bump et al. 1947).

Roosting

Ruffed grouse roost day or night on the ground and in trees. Commonly used winter sites include densely-branched conifers and under small conifers with low-hanging branches. Grouse will often roost beneath deep, powdery snow during periods of constant winds and cold temperatures. They bury themselves in snow by walking to a suitable spot and digging themselves in or by plunging in from the air—either directly from full flight or by dropping from a branch.

Grouse will often roost beneath deep, powdery snow during periods of constant winds and cold temperatures.

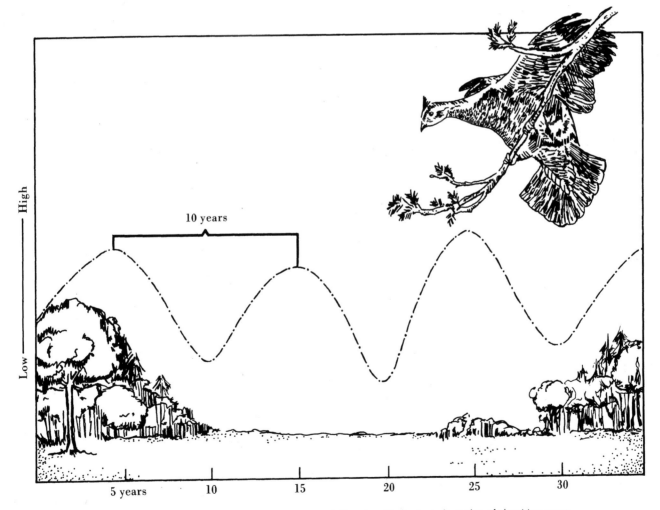

Figure 40. For reasons not fully understood, ruffed grouse populations tend to fluctuate in cycles of about ten years.

Abundance and Longevity

The quantity and quality of food and cover are probably the most important determinants of grouse abundance. Ample amounts of high-quality food and cover generally result in relatively small grouse territories and high numbers of grouse per unit area.

Adverse weather conditions can significantly affect the survival of grouse. Prolonged rains and cool temperatures in spring can be detrimental to young grouse. Insufficient snowfalls that prevent adults from roosting under snow during severe cold periods can also have unfavorable outcomes. Severe weather and low-quality food resources can weaken individuals and increase their susceptibility to predation and disease. Predators account for a large share of the overall mortality of grouse. Great horned owls, goshawks, Cooper's hawks, and red and gray foxes are the primary predators of adult grouse. Foxes, weasels, and skunks readily consume eggs, and Cooper's and sharp-shinned hawks prey upon chicks (Bump et al. 1947; Edminster 1947).

For reasons that are not well understood, grouse tend to undergo cyclic fluctuations of yearly abundance. Their numbers tend to be high for several years and then low for several more in a predictable cycle of approximately ten years (figure 40).

The **longevity** of ruffed grouse is typically short due to predation and other causes of mortality. Fewer than 50% of the chicks reach their first autumn (Bump et al. 1947). Of 1,000 chicks hatched, approximately 180 will survive to the following spring and sixteen may live to a fourth spring (Gullion 1981).

The longevity of ruffed grouse is typically short due to predation and other causes of mortality.

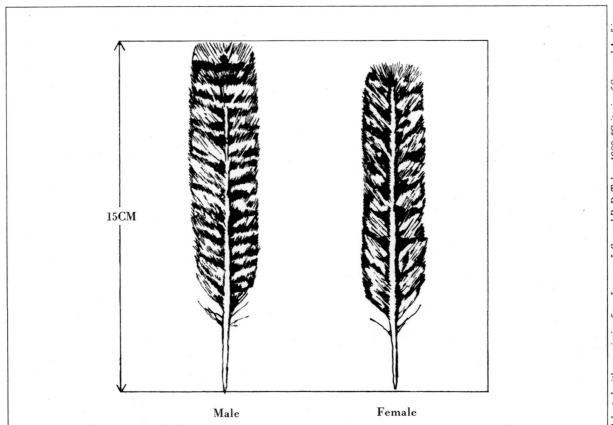

Figure 41. Central tail feathers of ruffed grouse may be used in determining sex.

External Sex Characteristics

Determining the sex of ruffed grouse externally can be difficult and often requires the examination of several characteristics. A continuous **subterminal band** on the tail usually designates a male, but if this band is interrupted at the two central tail feathers the bird may be either a male or female (figure 41). Males have longer ruff feathers than females and generally have an orange/red patch of skin above the eye. The front neck feathers of males tend to have bold, clearly defined light and dark bands, while those of females tend to be drabber and more blended.

Habitat Requirements

Ruffed grouse prefer forests that are heavily interspersed with areas in early stages of forest succession. They are not attracted to extensive areas of mature forest, because most of their food and cover plants are early successional species which die under a canopy of tall trees. Young woodlands, brushy lowlands, recently logged areas, overgrown pastures, and abandoned orchards provide essential habitats for drumming, courtship, nesting, feeding, and roosting.

Ruffed grouse prefer forests that are heavily interspersed with areas in early stages of forest succession.

Drumming and Courtship Sites

Drumming sites are typically located within brushy, open woodlands in intermediate stages of forest succession (figure 42). **Males**

Figure 42. Drumming sites are typically located within brushy open woodlands in intermediate stages of forest succession.

are strongly territorial; therefore, the drumming site and essential food and cover resources should be available within an area of 6–10 acres in size. The male usually selects a site that enables high visibility of surrounding ground for a distance no less than 60 feet (Gullion et al. 1962).

Downed logs, stumps, stonewalls, or boulders at least 8 inches in height are most commonly used for drumming. They are usually located within moderately dense brush and saplings, apparently to provide some protective screening against avian predators such as goshawks and great horned owls. The composition of the forest canopy above the drumming log varies, but many drumming logs are located where the canopy is relatively open.

Nesting and Brood-rearing Habitat

Female ruffed grouse sometimes nest more than ½ mile from the drumming site at which mating occurs (Gullion 1981). Preferred nesting sites include open woodlands with minimal undergrowth where the female can maintain unobstructed surveillance for predators. Nests are commonly located within 100 feet of woodland clearings such as recent clearcuts and herbaceous openings.

When the young hatch, the female leads them away from the nesting site and brings them into a denser understory. Broods require young, open woodlands or clearings with an abundance of herbaceous

Preferred nesting sites include open woodlands with minimal undergrowth where the female can maintain unobstructed surveillance for predators.

Figure 43. Clearings or sparse woodlands with an abundant herbaceous ground cover are preferred sites for brood rearing.

and young, woody plants. Tangled understories are initially avoided but may eventually be used as the brood matures. Favored brood-rearing sites include alder swales and sapling stands of aspens and other hardwoods (Gullion 1981). Broods commonly are found along woodland access roads, along abandoned **skid roads** and **logging headers**, and in other small clearings producing prolific young growth (figure 43).

Feeding Habitat

As is the case with woodcock, good feeding sites are perhaps the most critical habitat component for attracting and managing ruffed grouse. Early succession forests are crucial to adults and broods during the warmer months for providing an abundance of insects and fruiting plants. As these sites mature, the grouse food plants become shaded out by longer-lived tree species, and the site deteriorates as a feeding habitat. During colder months, the presence of various shrubs that retain their fruits well into winter and of shrubs and trees that offer palatable flower buds is very important.

Where present, aspen is a valuable component of the ruffed grouse food base. Aspen flower buds and catkins provide a nutritious high-energy winter and spring diet and are more accessible on individual, stout twigs than the buds of finer-branched trees such as birches, witch hazels, hornbeams, or other locally important grouse foods. In areas where aspen is not a significant component of the forest, the presence of the tree and shrub food sources becomes much more important.

Roosting Habitat

The requirements for roosting will generally be satisfied if mixed conifer and hardwood stands ranging from sapling to nearly mature stages are present. Preferable ground roosting sites are those with moderately dense brush and saplings, provided that an unobstructed view of nearby ground exists. Grouse often roost in conifers, mountain laurel, and mature aspens, particularly during the colder months. The presence of conifers is not essential if high-quality ground roosting sites are available. If not available, sparingly scattered clumps of low-growing conifers or mountain laurel serve well as winter cover during cold, windy periods. Tall, isolated conifers are not desirable because they are more exposed to winds and often provide avian predators ideal, year-round perches from which to prey on grouse.

Are Grouse Using Your Woodlot?

In addition to observing them directly, several clues are useful for detecting the presence of grouse. Drumming grouse may be heard year-round, but especially in spring and autumn. Downed logs larger than a foot in diameter that have a pile of droppings on and around them are a good indication that a male grouse has found a drumming log and established a territory (figure 44). Small, round, slightly dished areas of bare soil or sand may reveal a dust bath. Such areas appear to have been scratched and rolled in and may have a few loose feathers in them or scattered about their periphery. Large numbers or clusters of feathers elsewhere, however, indicate the presence not only of grouse

As is the case with woodcock, good feeding sites are perhaps the most critical habitat component for attracting and managing ruffed grouse.

but also of predators. In snow, tracks and snow roosts are telltale signs of grouse (figure 45). Vacant snow roosts appear as two neat holes in deep snow—one where the grouse entered and one nearby where it exited. If the observer carefully excavates the roosting site, the bird's droppings will usually be found inside. When one hole is seen from a distance, the grouse may still be inside and the observer might be in for a surprise!

Managing Habitats for Ruffed Grouse

Before implementing a management plan for ruffed grouse, it is important to become familiar with the property involved. By envisioning the habitat requirements of the species, a comprehensive idea of the management needs can be obtained. The following are some of the questions that might be considered when deciding "what needs to be done where." What does the property contain for food resources? Does it have an abundance of one food plant or does it contain various species in case the primary crop fails? Does it have an aspen component, and if so, how extensive and of what approximate age? Are large downed logs or boulders present? Is it fairly even-aged or is it interrupted by several sites in various stages of forest succession?

Adjacent neighbors may wish to coordinate efforts to manage a habitat if one property is smaller than the male grouse's territory or if each neighbor can contribute valuable habitat components.

Adjacent neighbors may wish to coordinate efforts to manage a habitat if one property is smaller than the male grouse's territory or if each neighbor can contribute valuable habitat components.

Figure 44. Droppings on a large downed log may indicate an established grouse drumming site.

Drumming Log Management

Drumming logs are usually not limiting factors to grouse abundance, but logs can be created if few are present. Several drumming logs at least 80 feet apart can be established by felling injured or poor-quality trees greater than 10 inches in diameter. Where possible, sites should be established either at the top or at the foot of small slopes so the drumming grouse attains maximum visibility in the event that predators or other ground should approach. Stumps should be left at least 3 feet high, and tree tops should either be removed or cut so that the logs lie flat on the ground. If large trees are not present, several small logs can be cut and stacked to heights of 12–16 inches. In either case, it is desirable to lay the drumming logs on the ground against a standing tree or tall stump. The tree or stump serves as a **guard object** and should be located not more than 8–10 feet away from the butt of the log (Gullion 1984).

Two of the most critical factors in creating drumming sites are the size and density of vegetation surrounding the log. Vegetation within a radius of 10 feet from the log should consist of moderately dense brush or saplings that are from 8 to 30 feet high (Gullion 1984). In hardwood stands possessing fewer stems, surrounding vegetation

Two of the most critical factors in creating drumming sites are the size and density of vegetation surrounding the log.

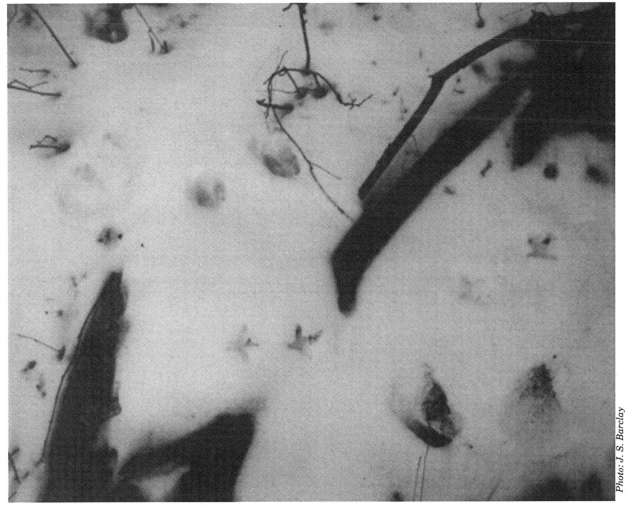

Figure 45. Tracks and snow roosts are telltale winter signs of grouse.

should be cut after leaf fall in autumn or in winter to generate sprouting of a greater number of stems in the spring. Any large trees to be cut should be felled away from the drumming log and preferably removed (perhaps as firewood) so that hiding places for predators are not created. The response of grouse to sites that have been cut may not be immediate, but may take several years depending upon the vegetation removed.

Ideally, drumming logs should be selected for rot resistance (species such as oak, hickory, elm, locust, and tupelo), and the surrounding vegetation should be selected for its ability to provide substantial brushy cover.

The type of woodland favored for ruffed grouse drumming sites can also serve as woodcock nesting cover when located near singing grounds. Also, the drier portions of woodcock feeding cover can be managed simultaneously for grouse by creating drumming logs if none are present.

Feeding, Nesting, and Brood-rearing Cover Management

Aspen is considered to be the most important ruffed grouse food and cover resource across much of the grouse's range. Many sites, especially in southern New England, lack a significant aspen component. Therefore, this section is divided according to the amount of aspen present and according to the landowner's willingness to manage for aspen.

Sites Lacking Aspen

Management for grouse should concentrate on maintaining several stands in different stages of forest succession. If various age classes are missing, strips or blocks of mature forest can be clear-cut to promote dense sapling growth. The clearcuts should be at least 1 acre in size (no greater than 10 acres), and a new strip or block should be cut every ten to fifteen years so that a rotational schedule of forty to sixty years is attained. Therefore, each patch will be cut only once in a forty- to sixty-year time span. Clear-cutting can be co-ordinated with a timber sale and/or fuelwood operation to reduce costs to maximize the utilization of forest products. If timber or fuelwood is desired, the rotational schedule can be adjusted accordingly as long as at least one area is cut every ten to fifteen years.

The clear-cutting operation should be as neat as is feasible. When possible, felled trees should either be removed from the site or reduced so that they lie flat on the ground to hasten rotting. By removing the logs and brush, more sunlight reaches the forest floor to encourage sprouting, and horizontal cover used by predators is reduced (Gullion 1984). Where deer feed heavily on new sprouts, however, loosely piling the brush over recently cut stumps may be necessary to somewhat deter the deer and increase the chances of successful sprouting (Lyons and Broderick 1986).

Clearing land and setting back forest succession creates brood habitats. Nesting habitats become available as the clearcuts approach the end of their rotational schedule and the dense undergrowth dies. Brood habitats for grouse can also be provided by some of the sites prepared for the management of woodcock feeding, nesting, and brood-rearing cover.

Aspen is considered to be the most important ruffed grouse food and cover resource across much of the grouse's range.

Desirable food plants should be encouraged wherever they are found. Abandoned apple orchards and patches of barberry are especially valuable as fall grouse resources. Any salvageable old apple orchards, scattered apple trees, and berry-producing trees and shrubs should be saved and *released* as necessary. When desired, food plants can be planted. It is generally best to transplant wild stock (saving as large a ball of soil around the roots as possible) when introducing new food plants to a site.

Sites With Only Scattered Aspen Clumps

On many sites aspen is sparsely distributed in small clumps, or *clones*. Management procedures for these areas are the same as for an area lacking aspen, but the aspen clumps should receive special care if the aspen component of the forest is to be expanded.

Aspens are short-lived, early successional species requiring full sunlight. When adequate sunlight reaches the forest floor, aspens send out a network of roots capable of sprouting numerous, rapidly growing suckers. Aspens usually live for only forty to fifty years, so if forest succession proceeds without disturbance, they eventually die in the shade of other trees. On sites where only scattered, mature clones remain, surrounding tree species can be clear-cut to reduce competition for a radius of approximately 70 feet or until the ground to the south receives full sunlight. Some of the aspens in the clone should be felled to encourage root sprouts.

Aspen clumps a few hundred feet in size can be managed as a group of four separate wedge-shaped units. One unit could be clear-cut every ten years starting with the southeast unit, then proceeding to the southwest, then to the northeast, and finally to the northwest unit. Clear-cutting should include a significant area surrounding the unit to allow for aspen expansion. The southern units are cleared first to allow maximum sunlight penetration into the stand and onto the cleared forest floor. If the initial aspen clump is old and in danger of dying, the cutting schedule can be hastened or more than one unit can be cleared simultaneously, but at least one unit should be left as a food source (paraphrased from Gullion 1984).

Sites Having a Prevalent Aspen Component

Where aspen is abundant, a clear-cutting schedule similar to that described for sites with no aspen can be followed. The period between clearcuts should be ten years instead of ten to fifteen years, because aspen grows more rapidly than other hardwoods. Therefore, each strip or block is cut once every forty years.

Aspen stands present in proper age classes fulfill the habitat requirements for feeding, nesting, brood rearing, and drumming. Saplings stands between four and fifteen years of age possess dense vertical cover attractive to broods, and stands between six and twenty-five years of age contain proper stem densities for drumming sites. Aspens over twenty-five years old are used heavily as food sources and fulfill the cover requirements for nesting and winter roosting (Gullion 1984).

> *Desirable food plants should be encouraged wherever they are found. Abandoned apple orchards and patches of barberry are especially valuable as fall grouse resources.*

Managing Roosting Sites

If the proper age classes of forests are present to satisfy drumming, nesting, brood rearing, and feeding needs, then sufficient year-round roosting cover should be available. However, grouse often roost in conifers during severe winter weather, especially when proper age classes of hardwoods are not available to provide cover. Conifers can be planted sparingly, but they should not be allowed to mature in, or adjacent to, grouse cover if avian predators are to be discouraged. Depending upon the manager's objectives, it may be desirable to remove tall conifers that are lightly scattered throughout grouse cover. Low-growing conifers can be maintained by a rotational schedule of planting, whereby small clumps of three to ten seedlings are planted when older conifers approach 10 feet in height. The tallest ones are then removed as the younger conifers become well-established.

Conifers can be planted sparingly, but they should not be allowed to mature in, or adjacent to, grouse cover if avian predators are to be discouraged.

Review Questions

1. What are the required site characteristics for woodcock singing grounds? Managing one singing ground per _____ – _____ acres of woodcock habitat is adequate.

2. What are the necessary components of a "drumming" site for ruffed grouse? One "drumming" site should be available for every _____ – _____ acres of grouse habitat.

3. What is the principal food of American woodcock? Based on this food source, when and why do woodcock migrate? What soil types and plant species characterize sites supporting this primary food source?

4. What general foods do adult and young ruffed grouse utilize in the summer? Which portions of some specific trees and shrubs are crucial to grouse survival in winter? Therefore, what forest conditions (e.g., successional stages) constitute a suitable summer feeding area? A suitable winter feeding area?

5. What forest conditions are necessary to satisfy nesting and roosting requirements for woodcock and grouse?

Field Exercises

1. Obtain a photocopy of the field map that you sketched for Chapter 2 and label it **Woodcock Habitat Map**. Review questions 1, 3, and 5 on the previous page. Then, either walk through your property or use existing map data to mark all areas appropriate for either singing grounds, nesting sites, feeding covers, or roosting fields.

    Examine your **Woodcock Habitat Map**.

    Are any of the four critical habitat types in short supply? If so, which ones?

    Mark any locations that contain at least two of the four required habitat types within a 5–10 acre area. These areas can be prioritized for woodcock habitat management.

2. Obtain a second photocopy of your original field map and label it **Grouse Habitat Map**. Review questions 2, 4, and 5 on the previous page. Then, either walk through your property or use existing map data to locate areas presently serving, or having potential to serve, as sites for "drumming," nesting, summer feeding, or winter feeding.

    If you walk through your property, can you find any drumming logs that have recently been used?

    Examine your **Grouse Habitat Map**.

    Which of the four required habitat types are in shortest supply, if any?

    Mark locations that contain at least two of the required habitat types within a 6–10 acre area. Do any of these overlap with the areas prioritized for woodcock management? Are there specific sites that satisfy some requirements for both woodcock and grouse?

These exercises were designed to be done on your own woodland. Once completed, you will be well on your way toward your own wildlife management plan.

Field Notes:

4 White-tailed Deer and Eastern Wild Turkey

Trends in the densities of white-tailed deer and eastern wild turkeys have been nearly opposite those for woodcock and ruffed grouse.

Introduction

The extensive cutting of northeastern forests near the start of the twentieth century provided an abundance of early successional habitats that supported relatively high densities of American woodcock and ruffed grouse. Then, as northeastern forests matured, early successional habitats diminished and woodcock and grouse densities declined.

Trends in the densities of white-tailed deer and eastern wild turkeys have been nearly opposite those for woodcock and ruffed grouse. Population densities of white-tailed deer and wild turkeys were extremely low at the turn of the century. Sound hunting regulations coupled with timber management practices and farm abandonment, however, have since contributed to dramatic increases in white-tailed deer populations. Restoration efforts in the northeastern United States since 1975 plus the maturation of New England forests have led to the reestablishment of steadily increasing and expanding wild turkey populations.

Chapter 4: White-tailed Deer and Eastern Wild Turkey

Many northeastern forests are harvestable and rapidly approaching overmaturity. Timber management practices, including harvesting and fuelwood cutting, can be highly compatible with wildlife management practices when tailored for particular sites. This chapter focuses on management strategies designed to retain and improve mature habitats while creating and maintaining patches in early succession. Some of the needs of white-tailed deer and wild turkeys can be coordinated readily with the creation and/or management of early successional stages for American woodcock and ruffed grouse.

White-tailed Deer

Description and Range

White-tailed deer in New England are technically known as northern woodland whitetails (*Odocoileus virginianus borealis*) of the family Cervidae. They are the largest in size and most widely distributed of the many white-tailed deer subspecies present in the United States.

Adult whitetails measure approximately 36 inches from ground to shoulder. Body weight is highly dependent upon factors such as sex, age, heredity, and habitat quality, but adult females generally weigh about 100 pounds while adult males average about 150 pounds. Males occasionally attain weights greater than 300 pounds.

Some of the needs of white-tailed deer and wild turkeys can be coordinated readily with the creation and/or management of early successional stages for American woodcock and ruffed grouse.

Figure 46. The white underside of the tail communicates alarm when the deer is threatened.

Deer shed their hair twice each year in accordance with seasonal needs. The summer coat appears reddish brown and consists of relatively short, solid hairs. By winter the coat appears bluish gray to dark brown and is comprised of large hollow hairs and sparse, wool-like underfur, which provide an effective means of insulation. The underside of a deer's tail is entirely white and communicates alarm when the deer feels threatened. The tail is twitched from side to side when the deer is nervous, fanned out conspicuously when the animal is moderately alarmed, and often raised and waved like a white flag when the deer flees from immediate danger (figure 46). The white hair on the tail is concealed when the deer is hiding.

White-tailed deer are found in all forty-eight contiguous states. The northern woodland whitetail is present throughout New England and ranges north to the Hudson Bay, west into Minnesota, and south to southern Illinois and Maryland (figure 47).

Life History

Antler Development

In preparation for the breeding season in autumn, male deer (bucks) develop antlers. Unlike the permanent horns of cattle, sheep, goats, and antelopes which consist of a **keratin** coating and bony core, antlers are entirely bony late in development and are shed and regrown annually. Antlers become visible in mid-March or April and continue to grow within a coating of "velvet" throughout the summer.

Unlike the permanent horns of cattle, sheep, goats, and antelopes . . . , antlers are entirely bony late in development and are shed and regrown annually.

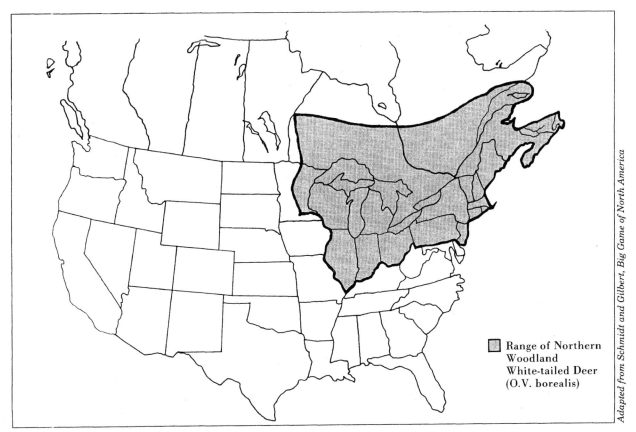

Figure 47. The northern woodland whitetail is the most widely distributed subspecies of deer in the United States.

The antlers are fragile while developing, and early injuries may result in deformities. By August or September, growth ceases, the antlers harden, and the velvet dries and sloughs off. The antlers are shed from mid-December through January (Behrend and McDowell 1967).

The size of antlers depends upon the deer's age and the quality of the habitat. Generally, they increase in size each successive year until the buck passes his prime (4½ to 7½ years) and are larger in areas with foods containing an abundance of essential nutrients, such as calcium and phosphorus. The shape and number of points may depend primarily on heredity (Rue 1978).

The Rut: Courtship and Fighting

The **rutting**, or breeding, season for white-tailed deer begins in September. The physical condition of bucks changes noticeably by autumn—antlers become well developed and muscle development is substantially increased, resulting in swollen necks and heavier body weights. Bucks become edgy and use their antlers to spar with saplings or other bucks. By sparring, young bucks assess their own strength and prepare for possible encounters during courtship.

Each female deer (doe) is in estrus (fertile and receptive to copulation) for approximately twenty-four hours of her twenty-eight-day estrus cycle during the courtship period in October through early January. When a buck senses that a doe is in estrus, he follows her and attempts to defend his rights to her to the exclusion of other males. After mating, the buck generally stays with the doe for several hours and continues to ward off competitors. Courtship peaks in mid-November, and an unmated doe may be in estrus several times throughout the courtship period.

Older bucks from the same area generally establish a hierarchy of dominance before the rutting season begins and avoid aggressive confrontations during the rut. When fighting occurs, it is usually between individuals of similar size—either dominant males of different groups or a dominant male and a younger challenger. Confrontations begin with direct stares and laid-back ears. These signals usually cause the weaker individual to turn away; but, if not, the dominant buck slowly approaches with lowered antlers and raised hair. If this gesture still fails to intimidate the challenger, the bucks may charge at each other from several feet away and push and twist forcefully with their antlers until the weaker retreats. Confrontations are usually brief, and injuries are relatively uncommon (Marchinton and Hirth 1984). Bucks that have not established dominance are generally capable of mating but may not have the opportunity to do so.

Fawning

Young deer (fawns) are commonly born in mid-June but may be born between May and July—approximately seven months after conception in autumn. Fawns are usually born forelegs first and are capable of standing and nursing within thirty minutes after birth. Does generally do not select specific habitats for bearing young, but they may return to the same vicinity year after year. In high-quality habitats, does may bear two offspring or infrequently three, but young does and does in habitats of lower quality generally bear only one.

By sparring, young bucks assess their own strength and prepare for possible encounters during courtship.

Fawns are rusty brown and speckled with white spots that help break up the body's outline when hiding or resting (figure 48). Except when nursing, they spend little time with the doe for their first two weeks and stay concealed among ground cover while the doe feeds elsewhere. She visits the fawns as little as possible which minimizes the chance of predators following her scent to the fawns (Marchinton and Hirth 1984). When two fawns are born, they generally remain concealed in different locations, again minimizing predator detection. As the fawns mature, they become increasingly active, siblings appear together, and by autumn, they remain with the doe constantly (Marchinton and Hirth 1984). Female fawns that have access to a high-quality diet may occasionally breed their first autumn; otherwise, they breed their second year.

Bedding

At night deer may **bed** (rest) in the herbaceous growth of open feeding fields or clearings, but during daylight hours they generally move to woodland cover and select locations that provide more concealment (such as dense brush). In winter, or stormy weather, bedding sites are selected to provide warmth and protection from winds and heavy rain or snow. Deer usually remain bedded through heavy snowstorms and allow the snow to accumulate on their backs. This behavior prevents abandoning the warm bed, and the layer of snow provides additional insulation from winds and cold temperatures. Deer occasionally return to the same bedding sites, especially in winter.

Except when nursing, [fawns] spend little time with the doe for their first two weeks and stay concealed among ground cover while the doe feeds elsewhere.

Figure 48. Fawns spend little time with the doe for the first two weeks, remaining concealed among ground cover while the doe feeds elsewhere.

Feeding

Twigs, leaves, buds, fruits, and nuts of trees and shrubs—as well as grasses, sedges, **forbs** (broad-leaved herbaceous plants), agricultural crops, ferns, mosses, lichens, mushrooms, aquatic plants, and even dry leaves—are included in the diet of deer. In winter, deer may fast to some extent, limiting their feeding primarily to **browse** (twigs, shoots, and buds of trees and shrubs) and to any remaining mast crops that might be available under snow (figure 49). By spring, the deer's nutritional reserves become depleted, and the deer begin feeding heavily on the abundance of new herbaceous growth and browse. Weaning fawns initially feed on forbs; then fawns and adults continue feeding on herbaceous plants throughout the summer, shifting between different species as deer preferences, palatability, availability, and the nutritional value of each plant change with the seasons and weather. In autumn, deer feed heavily on mast crops if available. When available, acorns and beechnuts may be eaten nearly to the exclusion of other foods. Apples are another favorite, and deer are commonly seen in orchards once the fruit begins to drop. Freshly fallen leaves of certain tree species are also readily consumed, particularly dogwoods and other species having red leaves with high contents of residual sugars.

Yarding and Winter Feeding

When winters become severe and snow accumulations become too deep to permit easy travel, deer congregate in well-sheltered wintering areas called **yards** in northern regions. Yarding areas typically

When winters become severe and snow accumulations become too deep to permit easy travel, deer congregate in well-sheltered wintering areas called yards in northern regions.

Figure 49. Winter feeding is often limited to browse and whatever mast crops remain available during snow cover.

are dense stands of conifers in or adjacent to swamp and lowlands. Yards inhibit deep snow accumulations on the ground, reduce wind penetration, and generally modify the effect of severe winter weather. Deer may be confined to yards for prolonged time periods during persistent snowfalls. When this occurs, food sources become critical.

Browse within the yarding area may be rapidly depleted depending upon the number of deer, the size of the yard, and the amount of browse present. When food is in short supply, deer feed on any browse within reach and may stuff themselves with foods of low nutritional value. Heavy browsing results in an obvious division of vegetation called a **browse line**; the area will be stripped of all palatable forage beneath approximately 6 feet and be untouched above 6 feet, about the maximum height deer can reach (figure 50). Malnutrition and death often result if deer are forced to remain in yards when this stage is reached. (It is possible for a deer to starve with a full stomach if the foods it consumes are of poor nutritional quality.)

Some Factors Affecting Abundance and Longevity

Poor or marginal habitats combined with severe winters can take a drastic toll on whitetails. Browse availability becomes especially important throughout the deer's home range in late winter when fat reserves become depleted, fawns are developing rapidly in pregnant does, and bucks begin antler development. If browse is unavailable or has been depleted, substantial mortality can occur

Deer may be confined to yards for prolonged time periods during persistent snowfalls, and when this occurs, food sources become critical.

Figure 50. Heavy feeding can result in a "browse line" approximately 6 feet high, below which all palatable forage has disappeared.

regardless of the duration of confinement in the yarding area (Weber pers. comm.). In addition to being stressful and possibly causing death due to malnutrition, such conditions can weaken deer and increase their susceptibility to predators, parasites, diseases, and exposure. Also, pregnant does that become malnourished may produce a reduced quantity of milk, resulting in death of fawns and lowered overall reproductive success for that year (Youatt et al. 1965).

Where locally present, coyotes, bobcats, and black bears are predators of whitetails of all ages—on healthy adults as well as on fawns; on does ready to give birth; and on adults weakened by injuries, malnutrition, parasites, or diseases. Domestic dogs (pet and feral) can account for a significant loss of whitetails (Mattfeld 1984; Rue 1978). Well-fed pets often run deer just as readily as feral dogs, and both are especially destructive as predators in deep snow with a light crust or when they find deeryards. Domestic dogs and coyotes occasionally run deer onto frozen ponds containing ice so thin that the deer falls through or ice so smooth that the deer is unable to stand and defend itself.

Due to some of the mortality factors listed above, plus collisions with vehicles, hunting, poaching (illegal hunting), and other factors, few wild whitetails reach their predicted life span of twelve years. Whitetails raised in captivity, however, may live for eighteen or more years (Rue 1978).

Pregnant does that become malnourished may produce a reduced quantity of milk, resulting in death of fawns and lowered overall reproductive success for that year.

Figure 51. The transition zone between field and forest can be an excellent feeding habitat.

Habitat Requirements

Habitat components that are most important to white-tailed deer include feeding and wintering areas. Rutting, fawning, and bedding generally are not associated with stringent site characteristics; so the whitetails' habitat requirements for these activities are more likely to be satisfied if feeding and wintering areas are available. Whitetails prefer forests that are interspersed with fields/clearings; stands of dense conifers; and stands of variously aged hardwoods containing plenty of mast-producing trees and accessible, high quality browse.

Feeding Habitat

During the warmer months, deer prefer areas that produce prolific amounts of herbaceous foods. Frequently visited sites include woodland clearings; areas recently logged; abandoned farmlands; gas and power line rights-of-way; pastures; hayfields; Christmas tree plantations; orchards; nurseries; and agricultural fields planted with crops such as corn, alfalfa, and clover. Such fields are also important for the value of their edges and associated ecotones (see chapter 2). Many ecotones contain an abundance of palatable sprouts, shrubs, and saplings (figure 51). Mature woodlands with open understories may also produce adequate herbaceous ground covers and browse.

Whitetails in the northeastern United States greatly increase their chances of winter survival if a sufficient quantity of mast is available in autumn. Wolf trees and stands of mast-producing species such as oak and beech are extremely beneficial. The presence of accessible browse is crucial during the winter months. Browse must be tall enough not to be buried in deep snow, yet it must be low enough that it remains within reach. Large amounts of browse are commonly found in early successional woodland areas and in areas that have grown up in sprouts after logging or fuelwood cutting, but this stage usually lasts only a few years.

Winter Cover

Ideal wintering/yarding areas should have the following two characteristics: cover dense enough to collect most of a snowfall before it reaches the ground and an abundance of low, woody growth within the yard and/or in accessible, relatively open areas immediately adjacent to the yards (Smith and Coggin 1984; Rue 1978). Yards with southern exposures generally provide the greatest benefit because they increase snowmelt and provide maximum warmth. Sites commonly used for yarding include dense stands of laurel and stands of mature conifers located on southern slopes; in well-protected upland areas; or in stream valleys, swamps, and other lowlands.

Are White-tailed Deer Using Your Woodlot?

The most obvious year-round indicators of deer presence are tracks and droppings (figure 52). When favorite routes are used repeatedly, tracks become so clustered that trails eventually develop. Trails are especially evident in snow, through dense brush, along

Whitetails prefer forests that are interspersed with fields/clearings; stands of dense conifers; and stands of variously aged hardwoods containing plenty of mast-producing trees and accessible, high-quality browse.

slopes, and around obstacles (figure 53). The shape of droppings can be useful for identifying what foods have been eaten and where they have been obtained. Pellet-like droppings often indicate that a deer has fed on dry foods such as browse, while loose, mounded droppings may indicate a diet of fruit or herbaceous plants, perhaps from an orchard or field (figure 54).

Occasionally, deer hair can be found on barbed-wire fences or on broken branches; it is recognized as being coarse, light in color, and sometimes brittle. A careful examination of browse may also reveal habitat use by deer. A deer does not have upper front teeth (incisors), so it breaks off browse by pressing the lower incisors against a hard pad on the roof of the mouth and jerking its head upward. Twigs that have been broken by deer have ragged edges and fibers as opposed to the neatly clipped edge made by a rabbit using upper and lower incisors.

In autumn, evidence of rutting behavior may be found. When bucks test their antlers and release aggression and frustration, they rub against saplings, leaving obvious marks, called **rubs**, approximately 2½ feet high where twigs have been broken and the tree has been debarked (figure 55). Bucks also leave **scrapes** along their trails, which may be used once or twice daily. Scrapes are made by pawing the earth bare, often directly below an overhanging twig that the deer has broken. At sites where two bucks have fought, the leaf litter or turf may be dug up and pushed aside, and occasionally small clumps of hair may be found.

Twigs that have been broken by deer have ragged edges and fibers as opposed to the neatly clipped edge made by a rabbit using upper and lower incisors.

Figure 52. Groups of pellet-shaped droppings like the one pictured are sure signs of deer presence.

In autumn and winter, areas where deer and wild turkeys have pawed through leaf litter and snow in search of remaining mast may be conspicuous. In winter, deer use of yards depends upon the particular region's climate. When deer are confined to yards, well-worn trails form in the snow as they move to feed, and browse lines often become obvious. Dropped antlers and signs of predation may also be found in winter.

Managing Habitats for White-tailed Deer

The seasonal home range of whitetails often encompasses several square miles. Many deer (especially does), however, spend much of their time in smaller areas, and all deer tend to frequent favorite sites if undisturbed. Managing an entire home range is prohibitive to most landowners, but a portion of the home range can be managed to provide or supplement one or more habitat requirements—namely feeding and wintering sites.

Feeding Habitat Management

Providing Herbaceous Vegetation

One of the simplest means of providing herbaceous foods is to maintain existing openings. Fields, woodland clearings, access roads, trails, logging headers, and skid roads can be mowed annually or at

Managing an entire home range is prohibitive to most landowners, but a portion of the home range can be managed to provide or supplement one or more habitat requirements—namely feeding and wintering sites.

Figure 53. Deer may use favored routes repeatedly until trails become evident.

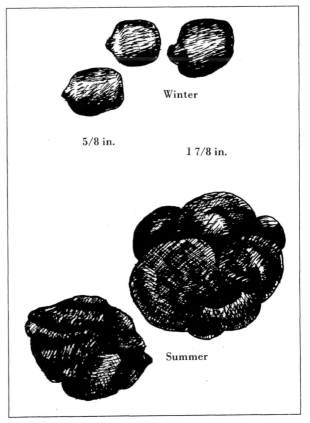

Figure 54. The shape of droppings can provide clues to what deer are feeding on.

least every three years to reduce the encroachment of woody species. These areas can be reseeded and/or fertilized if denser growth is desired. When sites are to be seeded, soils should be tested so that appropriate amounts of lime and possibly fertilizer can be applied. A general guideline for many sites might be lime spread at 2 tons per acre, and an early season mix of grass and legumes seeded at 25 pounds per acre (Weber 1986). Before seeding, it may be necessary to widen heavily shaded trails and roads by cutting adjacent trees. The increase in sunlight will stimulate herbaceous growth and the stump sprouts will provide browse (figure 56).

Where existing openings are lacking, it may be feasible to create patch cuts and maintain them as permanent herbaceous food sources. Frequent mowing (brush hogging), hand pruning, or herbicide application will likely be required to keep woody sprouts under control in areas targeted for permanent herbaceous openings.

Providing Browse

Browse can be provided most readily by cutting trees to encourage the growth of seedlings and stump sprouts. Timber harvests and fuelwood operations are especially important for providing browse in areas that otherwise might support extensive stands of trees too mature to produce accessible twigs, shoots, leaves, and buds. On sites lacking moderate amounts of browse due to the absence of tree harvesting or other openings, small patch cuts can be created.

Timber harvests and fuelwood operations are especially important for providing browse in areas that otherwise might support extensive stands of trees too mature to produce accessible twigs, shoots, leaves, and buds.

Figure 55. Bucks test their antlers and release aggression by rubbing against saplings.

Patch cuts should be less than 5 acres in size. Cuts larger than 1 acre should be elongated and irregular in shape. When possible, the patch cut should be created in hardwood stands that are highly preferred browse species (such as red maple), but not in stands that are in excellent condition for timber or as mast producers if such stands are in short supply.

A constant supply of browse can be achieved by maintaining openings or by coordinating efforts with timber and/or fuelwood production by rotating cuts so that as one area reaches its peak for browse production another opening is created. In areas that are to be maintained as browse, sprouts and saplings can be cut by chainsawing or mowing when most of the browse exceeds 7 feet in height. Desirable mast-producing shrubs that become established should be retained without cutting.

Browse is often in short supply when deer are confined to yarding areas in regions receiving snow accumulations greater than 18 inches. Immediate food sources can be provided if absolutely necessary by felling poor-quality hardwood trees (not conifers) or a preferred browse species. Deer often feed (usually at night) on felled trees and brush piles the same day they were cut. Such measures, however, temporarily boost the carrying capacity and possibly create negative long-term impacts by interfering with what might result in natural mortality.

A constant supply of browse can be achieved by maintaining openings or by coordinating efforts with timber and/or fuelwood production.

Figure 56. Removing large trees adjacent to woods, roads, and trails can stimulate the development of woody browse and herbaceous growth.

Chapter 4: White-tailed Deer and Eastern Wild Turkey

Providing Mast

Stands of mature mast-producing hardwoods should be encouraged. If timber harvests are planned, provision should be made to retain an adequate percentage of mature mast-producers—preferably oak and beech.

Healthy wolf trees of mast-producing species can be especially valuable for supplementing the autumn mast crop. They should be encouraged and retained unless they are suppressing the growth of trees and shrubs that might be more valuable as mast-producers or for timber production. Where possible, herbaceous vegetation can be encouraged beneath wolf trees to further increase their attractiveness as feeding centers for deer, grouse broods, and turkeys.

Managing Winter Cover

The lack of conifer stands suitable for winter shelter can be a major factor limiting the presence of deer in regions of the northeastern United States that receive snow accumulations greater than 18 inches. In regions receiving smaller snowfalls, the presence of conifers is important for providing immediate, temporary protection in heavy winter rains, snowstorms, winds, and cold temperatures.

Conifers greater than 5 inches DBH (diameter of the tree at breast height) are most useful as winter shelter. Smaller trees are not

Stands of mature mast-producing hardwoods should be encouraged. If timber harvests are planned, provision should be made to retain an adequate percentage of mature mast-producers—preferably oak and beech.

Figure 57. Undesirable trees such as maturing hardwoods in yarding areas can be killed by girdling. Such trees form snags which are beneficial to many wildlife species.

as useful because they tend to have flimsy branches inadequate for retaining enough snow to appreciably reduce ground accumulations. Cedars, hemlocks, spruces, and firs generally provide better cover than pines. Where a choice exists between spruce and fir (northern New England), spruce should be favored because it is longer lived and is less susceptible to deterioration by browsing deer, strong winds, and spruce budworm (Dickinson 1972; Weber pers. comm.). Management plans should allow for the perpetuation of stands suitable for yarding or immediate, temporary shelter. On sites lacking mature conifers, young conifers should be considered for their future value, and laurel stands should be retained.

Mature hardwoods competing in conifer stands designated for yarding should be judiciously eliminated to lessen competition for light, space, and nutrients. Hardwoods can be cut and removed or they can be killed by **girdling** or frilling. To girdle a tree, a chain saw or axe can be used to make a continuous cut around the trunk deep enough to cut through the inner bark and into the wood (figure 57). The cut should not be so deep, however, that the tree weakens and topples in a heavy breeze. When frilling a tree with an axe, a trough can be cut and herbicide can be poured in to ensure the tree's death. Girdled and frilled trees eventually form snags that are used by many wildlife species.

To encourage regeneration in conifer stands, consideration should be given to systematically harvesting some of the oldest trees as soon as commercially feasible. Stands should be cut lightly to guarantee that sufficient cover remains. Light cuts tend to favor regeneration of the conifers rather than invading hardwoods, and the resulting young conifers provide necessary concealment for deer.

Providing browse adjacent to winter cover or within extensive conifer stands is crucial when deep snows persist. Ideally, **corridors** of conifers should extend to areas with abundant browse so that travel lanes for easier winter access are available (figure 58). Browse can be managed as described above.

Providing browse adjacent to winter cover or within extensive conifer stands is crucial when deep snows persist.

Deer as a Nuisance

In many areas, whitetails have become so numerous that management plans focus on measures to control deer, rather than to encourage them. Overpopulated whitetails frequently feed on agricultural crops, fruit trees, nursery stock, Christmas trees, landscape plantings, and other foods outside of the deer's usual woodland habitats. They also may feed so heavily on the sprouts in a new patch cut that few sprouts are able to achieve more than a stunted development. When such feeding habits result in severe and costly damage, the whitetail's aesthetic benefits may be outweighed, and measures to discourage the nuisance deer may become necessary.

Control Measures for Timber Regeneration

In areas with large deer **herds**, cuts created to regenerate timber trees may receive such heavy browsing pressure that the only plants able to grow are likely to be unpalatable to deer and undesirable for future timber.

Patch Cut Shapes

Long, narrow patch cuts with meandering edges are generally most desirable for deer and other wildlife; the amount of edge is maximized and cover is nearby, even when an animal feeds in the center of the clearing. By cutting circular clearings several acres in size, however, deer feeding in the center lose the security of nearby cover. Although a large deer herd and lack of alternate food sources might induce deer to venture into the center of a circular cut, they are more reluctant to do so, and there is an increased chance that the sprouts in the center will survive (Lyons and Broderick 1986).

Brush Piles

Deer can be discouraged from certain areas of a patch cut by strategically piling the brush (*slash*) from the felled trees. Brush piled loosely over and around recently cut stumps will deter deer and allow a greater number of stump sprouts to grow free of browsing pressure. The brush piles also provide cover for other wildlife species.

Diversionary Patch Cuts

Where possible, a small patch cut can be created to serve as a diversion for a clearing designed to regenerate timber. It will not prevent deer from feeding on other sites, but it will likely absorb some feeding pressure to lessen browsing in the regeneration area. The diversionary patch cut should be located close to the clearing intended for regeneration and in a stand containing tree species that are highly

Where possible, a small patch cut can be created to serve as a diversion for a clearing designed to regenerate timber.

Figure 58. Ideal winter cover includes corridors of conifers which extend into areas of abundant browse.

preferred by deer. For example, if oak is the timber species to be regenerated, then the patch cut could be in red maple. The diversionary patch cut can be fertilized in its entirety or along its edges to further increase its attractiveness and palatability for deer. However, these cuts must not be so large or numerous that the carrying capacity for the site is boosted. Providing more quality feeding areas can lead to more deer and only compound the problem.

Control Measures for Farm Crops

The following discussion presents some available options for deer control. Detailed publications on preventing or reducing crop damage are available from county extension centers and state wildlife bureaus.

Fencing and Dwarf Varieties of Fruit Trees

Woven wire fences (such as livestock fencing) and electric high-tensile wire fences (such as the seven-strand slant wire and Penn State five-wire vertical fence) 6–10 feet high are effective for keeping deer away from crops. Unfortunately, constructing fence **exclosures** is expensive. Where fencing is not feasible to protect large orchards, prolific dwarf varieties of fruit trees can be planted to help defray costs. Although dwarf trees are more susceptible to deer browsing, higher density plantings are possible, fruit production per acre increases as tree density increases, and the installation of deer-proof fencing becomes more feasible (Caslick and Decker 1979).

Chicken wire or cattle fencing can be used to protect individual trees and shrubs. Also, the outside perimeter of fenced fields can be fertilized to encourage edge feeding so that deer are less inclined to attempt jumping 6- or 7-foot fences—causing possible injury, destruction of the fence, or further crop damage.

Repellents

A variety of repellents have been used with varying degrees of success. They should be selected for effectiveness, length of effective protection period, cost, and ease of application. A number of chemical taste repellents are available, and bars of soap, and/or bone meal, blood, and human hair hung in bags attached to young fruit trees throughout orchards may be helpful. Repellents are most effective when damage is first noticed. The longer deer have been feeding on crops, the more difficult it becomes to repel them.

Harvesting

Hunting is the most effective solution where deer are so overpopulated that they cause extensive crop damage and decreased yields. Proper harvesting of deer herds maintains a balance between deer numbers and present habitat conditions. Farmers incurring significant losses due to deer should consider allowing hunting on their property during the legal deer hunting seasons. In some states, farmers are given permission for year-round harvests if deer are a proven nuisance.

Electric high-tensile wire fences are effective for keeping deer away from crops or regenerating timber stands.

Eastern Wild Turkey

Description and Range

Several subspecies of the wild turkey exist in various regions of the United States. The subspecies found in the Northeast is the eastern wild turkey and is technically known as *Meleagris gallopavo silvestris* of the family Meleagrididae.

Adult turkeys are much larger than woodcock and ruffed grouse. They generally weigh more than 8 pounds and occasionally reach 25 pounds. Under poor lighting conditions, they appear black in color; but in full sunlight hues of iridescent bronze, brown, green, and purple become visible. A turkey's head is essentially bald, so the eyes appear large and the head long and narrow. Their long neck and long legs give them a streamlined appearance and make them well adapted for running up to 20 miles per hour. They are also capable of flying up to 55 miles per hour.

As is the case with ruffed grouse, wild turkeys are nonmigratory and are present *locally* in all six New England states. The eastern subspecies was originally distributed from New England to Florida and westward into the Plains States. Its range has been modified due to historic losses but is presently expanding due to changing land use practices and intensive restoration efforts during the past two decades (figure 59).

As is the case with ruffed grouse, wild turkeys are nonmigratory and are present locally in all six New England states.

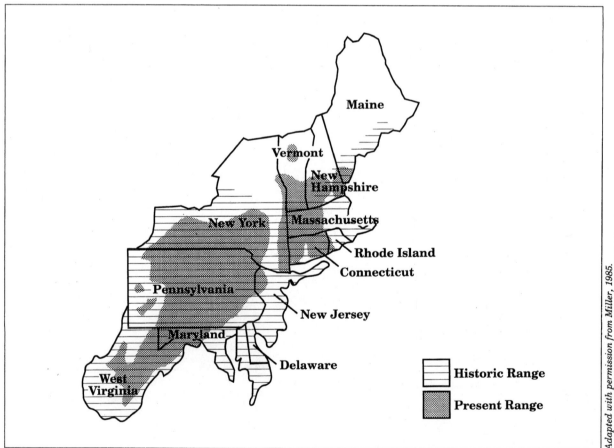

Figure 59. The range of wild turkey is expanding due to changing land use and restoration efforts.

Life History

Gobbling and Courtship

In the Northeast, courtship begins on warm mornings in late March, but may be hastened or delayed depending upon the favorability of spring temperatures. Male turkeys (toms or gobblers) advertise their presence to females and to other males by a vocalization known as *gobbling*. Gobbling males can be heard during the first hour after sunrise and first hour before sunset at the start of the courtship season and can be heard for increasingly prolonged periods as the breeding season progresses. Gobbling occurs occasionally throughout the year but is primarily restricted to courtship in the spring.

Males often gobble from their night roosts before flying to a nearby clearing in which to perform their courtship display. When at the display site, a dominant male gobbles and struts with tail fanned, feathers puffed up, head and neck pulled back against raised back feathers, and wings lowered to the ground (figure 60). Nearby females are attracted to the displaying male while subdominant males remain in small groups at the periphery of the site. This ritual may occur daily until the females leave in search of a nest site after mating, usually with the dominant male. As fewer females are attracted to his displays, his gobbling continues with increasing frequency and vigor until no more females are responsive or receptive to breeding. In New England, gobbling activity usually peaks in May.

The dominant male establishes a general territory and defends his display sites and his right to responding females to the exclusion of other males. Until dominance is established, fighting may occur frequently but is seldom fatal.

Nesting

Female wild turkeys (hens) begin nesting in April throughout the northeastern United States. The nest site is located under low-growing vegetation, beneath a tangle of vines (such as greenbriar or grape), or within the top of a fallen tree. In each location, the nest is frequently placed at the base of a tree or next to a fallen log. The female constructs a large cup-shaped depression in the leaf litter or grass with no assistance from the male. As with nesting woodcock and ruffed grouse, the protective coloration of her feathers blends with the forest floor and surrounding vegetation, concealing her and the nest from predators. When leaving the nest, females occasionally place leaves over the eggs to provide further concealment.

The female lays eight to fourteen eggs in a time span of ten to eighteen days. She spends less time visiting the dominant male as egg laying nears completion and severs all ties with him once a full clutch has been laid and incubation begins. The eggs are a uniform ivory to buff color often lightly speckled with rusty brown. They hatch after an incubation period of approximately twenty-eight days.

Unlike woodcock and ruffed grouse, **a nesting turkey may permanently abandon her nest after a single disturbance**. The closer she is to hatching her young, the less apt she is to leave. If a nesting hen is found, however, the observer should leave immediately and avoid repeated visits to the site for at least twenty-eight days (Miller 1985).

A nesting turkey may permanently abandon her nest after a single disturbance.

Chapter 4: White-tailed Deer and Eastern Wild Turkey

Figure 60. The displaying male gobbles and struts with tail fanned, feathers puffed, head back, and wings lowered to the ground.

Brood Rearing

Young turkeys (poults) are precocial—as are woodcock and grouse chicks. They are led away from the nest within hours after hatching, are able to fly short distances in one week, and become strong fliers by six weeks. Poults generally stay with the female throughout the summer and fall, during which time several hens with broods may join and form large flocks. If newly hatched turkeys are threatened by a predator, the female may pretend to have a broken wing in hopes of enticing the predator away from the nest.

Roosting

Turkeys younger than two weeks of age spend their nights on the ground under the fanned wings and tail of the female for warmth and protection. At approximately two weeks, the poults begin roosting 10–20 feet above ground in trees, and many initially sleep on the same branch under the female's outstretched wings. Poults roost higher and further apart as they mature, and by autumn, the family group may roost within an area up to 1 acre (Williams 1981).

Turkeys fly up to roosts at sundown and, depending on the availability of preferred roosting sites, may choose the same general area to roost in for several nights. Once a suitable limb has been selected and daily preening (cleaning) has been completed, they sit low, nestle their breast against the limb, and rest their head on their breast or back.

Turkeys fly up to roosts at sundown and, depending on the availability of preferred roosting sites, may choose the same general area to roost in for several nights.

Social Structure

In autumn, young males disperse from their family groups and join to form separate flocks (often with one or more adult gobblers) while family groups of adult females and their female offspring join and remain together until the start of the breeding season. Adult females that are unsuccessful at breeding, due to nest predation or other causes, usually join family flocks beginning in late August and remain with them until the following spring.

Older males are generally solitary after spring courtship and remain so throughout the summer. They form separate flocks in autumn (often with young males) and, except for occasional mingling at feeding sites, rarely associate with females until breeding season. Members in each flock establish a pecking order by fighting. The pecking order determines dominance within the flock and usually remains unchanged until the dominant male weakens or dies (Williams 1981).

Feeding

Wild turkeys eat a wide variety of foods as determined by the seasons and food availability. Poults feed heavily on succulent vegetation such as sedges and panicum grasses and on a great diversity of insects and other small animals such as grasshoppers, spiders, pillbugs, crickets, snails, worms, beetles, and fly larvae. In late winter and early spring, females feed on a wide variety of plant and animal matter to obtain protein and calcium necessary for egg laying and incubation. Through spring and summer, adults feed on insects, blades and seed heads of grasses and sedges, herbaceous plants, and berries (as they become available). Mast crops from trees such as oaks and beech are vital as autumn and winter foods. Other important autumn and winter foods include the catkins, berries, seeds, and buds of trees, shrubs, and vines such as black cherry, hop-hornbeam, ash, birches, pines, dogwoods, viburnums, barberries, sumacs, poison ivy, Virginia creeper, grapes, greenbriars, bittersweet, and honeysuckles.

In snow, turkeys scratch to uncover dormant insects, mosses, ferns (especially sensitive fern), tubers (from plants such as violets), and any remaining mast crops. When snow accumulations are deeper than 1 foot, they take advantage of southern slopes and areas where deer have pawed in search of food. Spring seeps are extremely important in winter for providing snow-free areas where turkeys can obtain more accessible foods. Waste silage and manure that are spread on snow on agricultural lands are also extremely important winter food sources and may be significant factors in the ability of turkeys to maintain and expand their northern range in New England (Blodgett pers. comm.; Miller pers. comm.).

Dominant males eat very little during peak periods of spring courtship due to an adaptation that allows them to concentrate on attracting hens. In winter, they accumulate a spongy layer of fat called a breast sponge from which they draw energy during courtship activities. The breast sponge initially weighs about two pounds and is usually depleted by summer (Lewis 1973; Williams 1981).

Spring seeps are extremely important in winter for providing snow-free areas where turkeys can obtain more accessible foods. Waste silage and manure that are spread on snow on agricultural lands are also extremely important winter food sources.

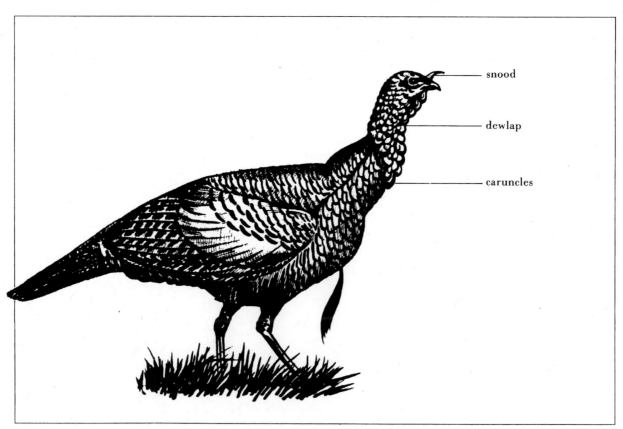

Figure 61. The male turkey is larger than the female, with a beard protruding from the breast; red, fleshy caruncles and dewlap on the head and neck; and a snood which hangs from above the bill.

Abundance and Longevity

As with any wildlife species, the quantity and quality of habitat resources are probably the most important determinants of the wild turkey's presence or abundance on a site. Sufficient high-quality food and cover resources usually result in healthy populations where the turkey is present.

Weather is an important factor affecting the success of broods and adults. Cold rains in the first two weeks of a poult's life can be fatal if the warmth of the female is not sufficient to dry the poult's feathers. Deep persistent snow accumulations hinder food searches and can result in death or temporary malnutrition. Also, late spring frost can lead to an inability of trees and shrubs to set seeds and leads to a reduction of mast crops during autumn and winter.

Predation can account for a significant share of turkey mortality. Healthy adults are susceptible to predation by bobcats; foxes; coyotes; and, occasionally, great horned owls and eagles. Adults weakened by malnutrition, disease, or injuries are especially vulnerable to various predators. Bobcats, raccoons, and great horned owls are among the few predators capable of preying on roosting turkeys (Lewis 1973). Primary predators of young turkeys include bobcats, foxes, coyotes, Cooper's hawks, sharp-shinned hawks, and owls. Crows, foxes, opossums, raccoons, skunks, weasels, mink, domestic dogs, and some snakes are among the egg-consuming predators.

The longevity of wild turkeys is typically greater than that for woodcock and ruffed grouse. Most turkeys fail to reach two years of age; but five-year-old birds are not uncommon, and some may live up to ten years (Lewis 1973).

External Sex Characteristics

Several characteristics are useful for sexing turkeys in the field. Males tend to be larger and possess black-tipped contour feathers (body feathers), beards protruding from their breasts, and red, fleshy **caruncles** and **dewlap** on the head—although the head coloration can quickly change from reddish to whitish blue depending upon the bird's mood (figure 61). Females are generally smaller than males, rarely have a beard, and have buff-tipped contour feathers. The backs of their heads are partially covered with small feathers and tend to be more drab in color than the male's head. Due to the color difference on the tips of the contour feathers, males appear almost black from a distance while females appear brown. Both sexes possess a **snood** on the forehead, but the male's is capable of elongating up to 6 inches by filling with blood during the peak of his display.

Tracks and droppings are also useful for determining sex. Males have large tracks 6–7 inches long while those of females are 4½–5 inches. Droppings of males are usually straight or J-shaped; and those of females are usually curled, corkscrewed, or mounded.

Habitat Requirements

Wild turkeys are benefitting greatly from the maturation of northeastern forests. They prefer extensive woodlands comprised of an abundance of mature mast-producing stands (such as oak and beech), conifers (such as white pine and hemlock) on up to 10% of the

Due to the color difference on the tips of the contour feathers, males appear almost black from a distance while females appear brown.

total area, and small forest clearings (including abandoned and active agricultural fields) comprising 10–40% of the total area.

Gobbling and Courtship Sites

The presence of sites suitable for courtship activities is important for attracting turkeys in the spring and for providing the population with opportunities to reproduce and increase in size. Displaying males require relatively open areas that have few obstructions and permit high visibility of the surrounding ground for approaching females or challenging males. Woodland access roads, field edges, small forest openings (resulting from fallen, wind-blown trees or other factors), and sections of mature hardwood forests with minimal undergrowth provide suitable courtship sites.

Nesting and Brood-rearing Habitat

Females often nest in the vicinity of the courtship area but may travel considerable distances when selecting a nest site. Although turkeys nest in a variety of areas, open woodlands with moderately dense undergrowth are generally preferred. Nests may be located near trails, access roads, or field edges and may be found in hayfields, fencerows, or power and gas line rights-of-way. The presence of open water near the nest site is evidently important because nesting females often fly directly to water before feeding (Healy 1981).

Displaying males require relatively open areas that have few obstructions and permit high visibility of the surrounding ground for approaching females or challenging males.

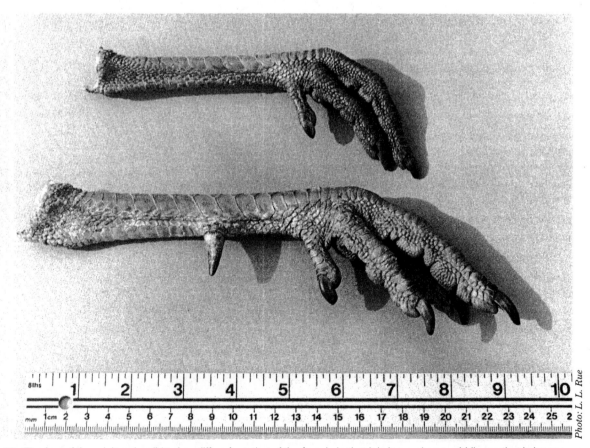

Figure 62. The foot of the adult male wild turkey differs from that of the female in that it is larger, has a middle toe that is longer than the other two lateral toes, and has a sharp spur above the foot.

Broods require open woodlands and clearings with herbaceous growth and an abundance of insects. Small patch cuts, recently logged areas, abandoned skid roads and logging headers, access roads, power and gas line rights-of-way, and stands of mature hardwoods with an herbaceous understory are highly favorable to succulent vegetation, insects, and turkey broods. Dense herbaceous growth is difficult for young turkeys to travel through and is avoided initially. By mid and late summer, they are able to feed in dense herbaceous ground covers as well as in woody understories (Healy 1981).

Feeding Habitat

The presence of hard and soft mast-producing trees and shrubs is vital to the existence of wild turkeys. Large, mature stands of hardwoods (oak, hickory, beech) and wolf trees along forest openings generally produce large annual crops of hard mast. Edges of forest openings favor soft mast-producing species such as dogwoods, viburnums, and grapes. Proper interspersion of areas in young and mature successional stages will fulfill the year-round feeding requirements of adults and broods.

Roosting Habitat

Wild turkeys select the tallest trees on a site for roosting. They prefer conifers that are sheltered from winds and located near clearings or open ridges—apparently for easier access to and from the treetops (Lewis 1973). Favorite roosting trees often overhang small woodland ponds and wetlands (Maciejowski pers. comm.). Clumps of large white pines or hemlocks usually satisfy their roosting requirements.

Are Wild Turkeys Using Your Woodlot?

The presence of turkeys can be determined by several means. Males can be heard gobbling throughout the spring—primarily just after sunrise and just before sunset. Loose feathers provide excellent positive identification and are commonly found during the fall **molting** period and in small forest openings used by males for displaying and fighting. Tracks and droppings are also excellent indicators (figure 62). Turkey tracks and droppings are large, and again, the larger tracks and straight droppings belong to males, while smaller tracks and curled droppings designate females. In spring, large tracks with continuous lines on either side reveal that a male has been displaying, as indicated by the drag marks of lowered wings. Turkeys use dust baths which appear as large, round, slightly dished areas of bare soil in the leaf litter.

Managing Habitats for Eastern Wild Turkeys

Wild turkeys occupy home ranges of 1 square mile or more, so an ideal management area (management unit) should be at least 600 acres and preferably 1,000 acres—an area larger than most private land holdings. Wild turkeys travel extensively within their home range, especially when they move from winter feeding areas to nesting sites and from brood-rearing cover to fall feeding areas. Therefore,

Wild turkeys occupy home ranges of 1 square mile or more, so an ideal management area (management unit) should be at least 600 acres and preferably 1,000 acres—an area larger than most private land holdings.

management of small properties can contribute substantially to the fulfillment of one or more of these seasonal habitat requirements, and a network of small properties managed throughout the turkeys' home range is especially effective. Where possible, efforts should be coordinated with adjacent landowners to assemble as much of a management unit as possible.

Courtship, Feeding, Nesting, and Brood-rearing Cover Management

Stands of Mature Hardwoods

Promoting stands of mature mast-producing hardwoods is extremely important to wild turkeys. Management plans should favor the retention of at least 50% of the total area in forest cover, with ½ of this percentage in mature hardwoods. Mature mast-producing trees should comprise at least 10% of the forest cover. In addition to supplying essential fall and winter food crops, mature stands provide brood-rearing cover. As the stands mature and thin naturally, herbaceous growth develops in response to greater light penetration to the forest floor.

Timber management practices, including harvesting and fuelwood cutting, can enhance turkey habitats. Harvesting creates feeding sites for broods and provides tree tops under which to nest (Miller 1985). Timber-stand-improvement operations generally encourage the healthiest trees while fuelwood operations often cull trees in poorer condition which are not valuable as sawlogs. Therefore, good mast-producing trees are encouraged.

Management of small properties can contribute substantially to the fulfillment of one or more of these seasonal habitat requirements, and a network of small properties managed throughout the turkeys' home range is especially effective.

Figure 63. Wild turkey poults, such as these 3½-month-old youngsters, feed avidly on insects found on herbaceous plants in forest openings and along field borders.

If timber harvests are planned, several guidelines for the benefit of turkeys might be considered. Mast-producing trees greater than 14 inches DBH (diameter of the tree at breast height) generally produce the largest mast crops (Miller 1985), so a high percentage of these trees should be retained unless a clearcut is planned. Clearcuts extending over many acres hold little value for wild turkeys, and any large cuts should be long and narrow instead of square or round (Williams 1981). Additionally, timber should be harvested by a method that insures healthy regeneration of a new hardwood stand (Mosby and Handley 1943).

Forest Clearings

Small forest clearings provide display sites for males, nesting sites for females, and feeding sites for adults and broods (figure 63). If no, or few, clearings are present within an area of 50 or more acres, then one or two (½–1 acre) or several smaller clearings can be created.

Clearings generally should not be created near roads and houses or along boundary lines. They also should not be created at the expense of prime mast-producing stands if such trees are in short supply. Stands having poor or marginal value for mast or other desirable attributes should be considered first. Turkeys are reluctant to stray far from protective cover when using clearings (Grenon 1986; Lewis 1973), so clearings from ¼ to 5 acres in size are most beneficial.

New and well-established clearings require periodic maintenance if they are to be kept at a stage suitable for courtship, summer feeding, nesting, and brood rearing. They should be kept free of extremely dense ground covers and maintained by mowing (brush hogging) every one to three years or by bulldozing. A strip at least 20 feet wide surrounding clearings should be mowed less frequently to provide nesting habitat and protective cover for young broods. Clearings and isolated fields can be mowed from the center outward to relocate and temporarily increase the densities of insects in the unmowed strip for broods (Hurst and Owen 1980). All clearings and isolated hayfields should be mowed after June 20. Delayed mowing prevents nest destruction and provides better feeding habitat for broods (Miller 1985).

Grazing cows, sheep, and other livestock can also be an important means of maintaining clearings. Light to moderate grazing maintains a relatively sparse ground cover favorable to turkeys and interrupts forest succession by hindering the natural development of fields to forest. Overgrazing, however, destroys suitable nesting cover; reduces turkey food plants and insects; and, therefore, eliminates potential feeding sites.

Supplemental Feeding and Food Plots

One may be inclined to supplement the diet of wild turkeys to "help" them through the winter and to attract them to an area where they can easily be observed. However, supplemental feeding not only is costly but can be more detrimental than beneficial. It can facilitate the transmission of diseases and parasites, attract predators by concentrating the turkeys at predictable times, and increase their vulnerability to poaching by taming them and increasing their dependence on artificial foods. Supplemental feeding is unnecessary and should be avoided at all times (Miller 1985; Williams 1981).

> *One may be inclined to supplement the diet of wild turkeys to "help" them through the winter and to attract them to an area where they can easily be observed. However, supplemental feeding not only is costly but can be more detrimental than beneficial.*

A better method to attract more turkeys and to sustain their populations is to improve the quantity and quality of food resources within their habitats. When forest clearings are created, plant diversity increases and shrubs and vines producing soft mast become established along the edges. Mast-producing species should be retained and encouraged by release cutting when necessary. Desirable trees, shrubs, and vines that are not present can be introduced by seeding or transplanting. Turkeys derive more benefits from mast species that retain their fruits well into winter than from occasional supplemental feedings.

Small clearings, or food plots, can be planted to provide year-round food sources. Food plots should be no larger than 5 acres but large enough so that other wildlife species can use them without depleting the turkey's food source. Species to be cultivated should be selected for site suitability and for the date at which they are most useful to turkeys. It is desirable to maintain several food plots that mature at different times so turkeys derive maximum year-round benefits. Crops favored by turkeys include rye, wheat, millets, corn, clovers, and peas. Ideally, the borders of food plots can be managed to encourage soft mast produced by shrubs and vines such as dogwoods, Japanese barberry, blueberries, honeysuckle, winterberry, and grapes.

In locations where deer tend to browse heavily, alternative crops or fence exclosures may be necessary to prevent food plots from "disappearing." Fencing is expensive but can be effective. If fencing is installed, a wooden gate or horizontal board at the top of the fence should be included. Turkeys are often reluctant to pass over a fence, but will jump up on a rail ("jump pole") to enter the food plot (Williams 1981).

Woodland access roads, trail edges, and abandoned skid roads and logging headers can be seeded with clovers, legumes, and grasses to provide summer feeding opportunities for adults and broods. Unharvested field corn provides an additional winter food source during periods of prolonged snow accumulation if mast crops are in short supply. Leaving a few rows of corn along an edge adjacent to forested land is valuable not only to wild turkeys, but to other wildlife species as well (Miller 1985).

Spring Seeps

Spring seeps are an important resource to include in the winter food base because they may remain unfrozen long after other areas have become frozen and snow covered. Seeps located on gentle, southern slopes remain open for longer periods of time and contain a greater abundance of herbaceous growth. The overstory ideally should be comprised of mature mast-producing hardwoods. Surrounding hardwoods can be thinned to encourage the regeneration of soft and hard mast-producing species whose fruits may lie dormant in the seep for months. Conifers directly over or immediately to the south of spring seeps shade the area in winter and prevent the development of succulent growth (Healy 1981). Conifers and dense brush that shade seeps should be removed to allow greater light penetration for increased herbaceous ground cover. Seep edges can be seeded with clovers and other low-growing crops to increase their attractiveness to turkeys.

Spring seeps are an important resource to include in the winter food base because they may remain unfrozen long after other areas have become frozen and snow covered.

Control of Domestic Dogs

The domestic dog is an important predator of wild turkeys in some locations (Grenon 1986). In addition to running deer and chasing other wildlife species, domestic and feral dogs have been responsible for a significant portion of nest losses and overall mortality on nesting females and on females raising young poults (Speake et al. 1985; Speake 1980). Dogs should be restrained so they do not range freely in areas inhabited by wild turkeys.

Managing Roosting Sites

Creating additional roosting sites is probably unnecessary if large trees are present (especially adjacent to small clearings). However, if roosting sites are limited, larger trees should be encouraged, i.e., clumps of five or six large conifers (white pine or hemlock) should be maintained where possible. When clearing land, provision should be made to save stands containing large groups of mature hardwoods and conifers if few are present in the immediate vicinity.

> *When clearing land, provision should be made to save stands containing large groups of mature hardwoods and conifers if few are present in the immediate vicinity.*

Chapter 4: White-tailed Deer and Eastern Wild Turkey

Review Questions

1. What are some important autumn and winter foods for white-tailed deer? Spring and summer foods?

2. What are deer yarding areas and under what conditions are they most crucial? What type of forest cover is best suited for deer yards?

3. What are some important autumn and winter foods for wild turkeys? What are some important spring and summer foods for adults? For poults?

4. What are spring seeps and why are they important to wild turkeys?

5. A well-managed turkey habitat might include conifers on up to _____% of the total area, small forest clearings on _____% of the total area, and an abundance of _____ on the remainder.

Field Exercises

1. Using range maps in field guides or other references, make a list of mast-producing trees and shrubs that grow in your region and which you believe are utilized by deer and wild turkeys. Visit your woodlot with a photocopy of the map that you sketched in Chapters 1 and 2 and estimate the percentage of your land that is dominated by these mast-producing species.

    Have your service or consulting forester assist you in determining the age and condition of these stands.

    How can the stands be improved for increased long-term mast production?

2. Label the above photocopy **Deer and Turkey Habitat Map**. Delineate any mast-producing stands or thickets that are present, and label them by tree or shrub species.

    Mark any small clearings, such as fields, clearcuts, utility rights-of-way, and vegetated access roads or trails.

    Label areas having well-protected stands of conifers or mountain laurel. If none are present, do suitable sites exist on adjacent ownerships?

    Pinpoint any stands or groups of large conifers or hardwoods that might serve as roosting sites for wild turkeys.

    Note which habitat components are absent or which ones need to be improved (if any) in order to enhance your property for deer or turkey.

3. Contact your Cooperative Extension educator, local wildlife biologist, or service or consulting forester to learn what the density of the deer population is in your area.

4. Look for signs of deer use on your woodlot. Which of the following signs are present?
    - _____ tracks
    - _____ trails
    - _____ browsed twigs, sprouts, or herbaceous plants
    - _____ droppings
    - _____ rubs
    - _____ hair
    - _____ scrapes

These exercises were designed to be done on your own woodland. Once completed, you will be well on your way toward your own wildlife management plan.

Field Notes:

5 Other Upland Forest Wildlife Species

In relatively uniform forests, as are found over much of the Northeast, timber and wildlife management practices can provide effective means for interrupting succession and increasing both habitat and wildlife diversity.

Introduction

The maturation of forests in the United States has resulted in a predominance of uniformly aged stands. Habitat diversity is often lacking, and only certain wildlife species are able to thrive as a result. Many wildlife species are associated with specific stages of forest succession, i.e., with the particular stem densities, canopy development, insect populations, and other characteristics unique to certain developmental stages. In relatively uniform forests, as are found over much of the Northeast, timber and wildlife management practices can provide effective means for interrupting succession and increasing both habitat and wildlife diversity.

Chapter 5 presents management practices designed to improve or maintain habitats for other upland forest wildlife. Important habitat components and their management in young, intermediate, and mature forests are discussed. Most of the management practices and habitat components addressed can be adapted to management plans for American woodcock, ruffed grouse, white-tailed deer, and eastern wild turkey with little modification.

Selected Species Groups

Rabbits and Hares

Four wild species of rabbits and hares are present in New England (excluding coastal islands)—the eastern cottontail, New England cottontail, snowshoe hare, and European hare (figure 64). Both cottontails are present throughout New England except in northeastern sections. The snowshoe hare is found across much of New England, while the European hare is found only in certain portions of New England. The eastern cottontail and European hare were both introduced in New England prior to 1910 (Godin 1977).

Rabbits and hares are herbivores which feed primarily on herbaceous vegetation in summer and on bark, twigs, and buds in winter. They are an important food base for predators such as bobcats, lynx, foxes, coyotes, and larger species of hawks and owls.

Special Habitat Requirements

Cottontails and hares require brushy fields and other forest openings that provide tender herbaceous growth for summer feeding, as well as sprouts, saplings, seedlings, and low branches for winter feeding. A sufficient amount of woody browse high enough to remain accessible after deep snowfalls is crucial to winter survival. Some favored woody foods in winter include red maple, apple, alder, aspen, birch, oak, white cedar, and balsam fir.

Cottontails and hares require brushy fields and other forest openings that provide tender herbaceous growth for summer feeding, as well as sprouts, saplings, seedlings, and low branches for winter feeding.

Figure 64. Available woody browse just above winter snow lines is crucial to the survival of rabbits and hares.

Photo: USDA Soil Conservation Service

Sufficient cover must be available at all times of the year, particularly after leaf fall and in deep snow. Stonewalls, brush piles, low conifers, and tangles of multiflora rose, raspberry canes, greenbriars, young alders, dogwoods, or spirea often serve as valuable resting and escape cover throughout the year. The height of cover is especially critical for maintaining snowshoe hare populations. At least 30% of an area managed for snowshoe hare should be composed of conifers less than 15 feet in height to provide dense, low branches for cover and food, especially in deep snows (Royar 1986).

Tree Squirrels and Chipmunks

Four species of tree squirrels and one chipmunk species occur in New England—the eastern gray squirrel, red squirrel, northern flying squirrel, southern flying squirrel, and eastern chipmunk (figure 65). The red squirrel and eastern chipmunk are distributed throughout the Northeast, the gray squirrel and southern flying squirrel are absent in northern New England, and the northern flying squirrel is absent along coastal southern New England (DeGraaf and Rudis 1986).

Tree squirrels and chipmunks feed on hard mast, soft mast, leaf buds, flower buds, bulbs, and mushrooms. They sometimes assume carnivorous feeding habits by consuming bird eggs, nestlings, and insects. Tree squirrels and chipmunks are commonly preyed upon by foxes, coyotes, and some hawks and owls.

Tree squirrels and chipmunks feed on hard mast, soft mast, leaf buds, flower buds, bulbs, and mushrooms. They sometimes assume carnivorous feeding habits by consuming bird eggs, nestlings, and insects.

Figure 65. Tree squirrels are opportunistic feeders known to consume eggs, insects, and even nestling birds along with hard and soft mast.

Special Habitat Requirements

Mature stands of hard mast-producing species such as oak, hickory, and beech are essential to gray squirrels and are favored by flying squirrels and chipmunks. Mature conifers are favorable to red squirrels, and all four squirrels and the eastern chipmunk commonly inhabit mixed deciduous and coniferous forests.

Although squirrels build leaf nests, tree cavities and abandoned woodpecker holes are preferred for raising young and winter denning. Chipmunks excavate networks of tunnels in leaf litter and loose soil, commonly in close association with stonewalls, rock piles, old stumps, and hollow logs.

Woodchucks

Woodchucks generally are associated with active agricultural lands such as hayfields and pastures, but they often live along forest edges, in large forest clearings, and in fields reverting to brush and trees. They belong to the squirrel family and are found throughout the Northeast.

Favorite foods include succulent grasses, alfalfa, clovers, and the tender shoots of other herbaceous plants. Woodchucks excavate large, complex burrows and hibernate through the coldest months. When abandoned, the burrows are an important source of cover for rabbits, foxes, skunks, weasels, opossums, and many other animals. Burrows are occasionally shared with other mammals in winter.

Special Habitat Requirements

Woodchucks thrive in cultivated or early successional areas that produce an abundance of succulent herbaceous vegetation. They usually are overlooked in management plans but appear as a result of habitat modifications for other species. Herbaceous openings managed for wild turkeys, for example, often attract woodchucks as well.

Small Mammals

Approximately five species of mice, four species of voles, two species of bog lemmings, seven species of shrews, three species of moles, and nine species of bats inhabit all, or parts, of New England. One bat species, the Indiana myotis, is endangered in New England (figure 66). It has not been observed in Connecticut since 1940 but has been observed more recently in Vermont and Massachusetts (Dubos pers. comm.; DeGraaf and Rudis 1986).

Mice feed regularly on both plant matter and *invertebrates*. Voles and bog lemmings feed on seeds, fruits, tender herbaceous vegetation, mushrooms, and occasionally on insects and other animal matter. Shrews, moles, and bats are primarily insect eaters (insectivores), but shrews and moles are *opportunistic feeders* which consume almost any animal small enough to capture. Small mammals provide a large and important food base for predatory birds, mammals, and snakes.

> *Mature stands of hard mast-producing species such as oak, hickory, and beech are essential to gray squirrels and are favored by flying squirrels and chipmunks.*

Special Habitat Requirements

Each small mammal species has specific habitat needs, but an abundance of low ground cover is generally required by all except bats. Dense herbaceous ground covers, tangles of brush, and downed trees and tops provide cover and food. Thick leaf litters, rotting logs, and stumps provide cover and abundant sources of millipedes, centipedes, beetles, snails, and other plant-decaying invertebrates. Spring seeps, brooks, and other wet areas are also important habitat components. Some streams and ponds are important feeding sites for bats. They tend to be free of overhead obstructions and often support large concentrations of emerging insects such as midges, mosquitoes, mayflies, caddisflies, and stoneflies.

Bats require well-sheltered areas for cover. Hollow trees, decaying snags with large plates of loose bark, rock outcrops, ledges, and caves are the most commonly used natural roosting sites. Six bat species remain in the northeastern United States throughout the year and congregate in favorite caves known as **hibernacula** for the winter. Three species—the silver-haired bat, red bat, and hoary bat—migrate to warmer southern states for the coldest months (Dubos pers. comm.; Godin 1977).

Bats require well-sheltered areas for cover.

Figure 66. Bats require well-sheltered areas for cover, such as hollow trees, ledges, or caves.

Canids

Three wild members of the dog family (Canidae) inhabit New England—the red fox, gray fox, and coyote. Red foxes inhabit both farm and forest land and are distributed throughout New England. Gray foxes typically inhabit forests and are present in all of New England except northern Maine. Coyotes are rapidly expanding their range and are now found virtually throughout New England.

The three canids are primarily predatory and often occupy the highest trophic levels. Their population densities are largely dependent upon the abundance of small mammals, rabbits, squirrels, and other prey. They will consume a wide variety of foods when available or when prey is scarce, at times becoming opportunistic and omnivorous. Other foods include birds, eggs, frogs, turtles, snakes, carrion, insects, and fruits. Unlike the other canids, gray foxes are capable of climbing trees to some extent to find cover and food (figure 67).

Special Habitat Requirements

One of the most important requirements is the presence of suitable habitat for preferred prey. Suitable habitats include woodlands that are interspersed with herbaceous clearings, edges, and early successional sites that have ample ground cover for small mammals, rabbits, and hares. Although canids can excavate their own dens, existing

Population densities [of canids] are largely dependent upon the abundance of small mammals, rabbits, squirrels, and other prey.

Figure 67. The fox and coyote are predatory mammals. A good habitat for small mammals is a good habitat for these canids.

Photo: USDA Soil Conservation Service

burrows are adopted readily. Abandoned woodchuck burrows generally fulfill the denning needs of red foxes; large tree cavities, hollow logs, hollow stumps, or protected rock outcrops suffice for gray foxes. Ledges and rock outcrops that form caves are often used by coyotes, especially for winter denning.

Felids

The bobcat and lynx are the only two officially recognized species of wildcats (family Felidae) that inhabit New England (figure 68). Many sightings of mountain lions (cougars/catamounts/panthers/pumas) have been reported but not confirmed. The bobcat is distributed throughout the Northeast, and the lynx is present only in northern New England.

Bobcats and lynx are strictly predatory and occupy the highest trophic levels in a food chain, web, or pyramid. Foods include rabbits, hares, squirrels, chipmunks, small mammals, birds, and fresh carrion. Lynx are so highly dependent on snowshoe hares that their population densities often mirror fluctuations in snowshoe hare densities.

Bobcats and lynx are strictly predatory and occupy the highest trophic levels in a food chain, web, or pyramid.

Special Habitat Requirements

Lynx are secretive in their habits and require large tracts of undisturbed forest. Bobcats are also secretive but tolerate human habitation better than the lynx, and will inhabit forests that are interspersed

Figure 68. Bobcats are secretive predators that occupy the highest trophic levels in the food chain.

with fields and agricultural land. Dense brush, wooded swamps, and young conifer stands are important to both species for cover year-round. Hollow logs, downed trees, and recesses and crevices in rock outcrops and ledges are important for denning. As with the canids, one of the most important habitat components is the presence of sufficient cover and food to support an abundance of prey species.

"Upland" Mustelids

The ermine (short-tailed weasel), long-tailed weasel, marten, fisher, and striped skunk belong to the family Mustelidae—so named because each possesses glands which emit a strong musk. The term *upland* is used arbitrarily to separate these species from the more aquatic mustelids, i.e., mink and river otter. Both weasels and the striped skunk are distributed throughout the northeast (figure 69). The marten and fisher are absent in much of southern New England, but their ranges appear to be expanding and now include portions of Connecticut.

Excluding the striped skunk, "upland" mustelids are largely predatory. Foods include rabbits, hares, squirrels, chipmunks, small mammals, birds, bird eggs, salamanders, frogs, toads, snakes, turtles, insects, and (occasionally) fruits and carrion. A large part of the fisher's diet often consists of porcupines. Skunks are omnivores and feed on insects, small mammals, bird eggs, seeds, fruits, buds, greens, and carrion.

Excluding the striped skunk, "upland" mustelids are largely predatory.

Figure 69. Weasels prefer early successional forests with an abundance of clearings, fields, and brushy edges.

Special Habitat Requirements

The weasels and striped skunks prefer early successional forests with an abundance of clearings, small agricultural fields, and brushy edges. The marten and fisher require large tracts of relatively mature forest interspersed with streams and other wet areas and are generally associated with conifers. Hollow logs, hollow stumps, large tree cavities, unoccupied ground burrows, and abandoned porcupine dens are important as denning sites.

Songbirds and Woodpeckers

Most songbirds belong to a large group of birds known as passerines, or perching birds. There are approximately 130 species of passerines and nine species of woodpeckers that breed in the Northeast. Some species are year-round residents and others only breed here and migrate to southern regions for the remainder of the year. Each species has a specific distribution which can be determined by consulting the range maps in most field guides.

Most woodpeckers feed on grubs and other insects found on dead or decaying trees, but fruits and nuts are also consumed by some species. Passerines feed on a variety of foods including seeds, fruits, insects, and other invertebrates. Each species occupies a specific niche which separates its feeding habits from those of other species. Such specialization tends to reduce competition and allows many species to inhabit the same forest. Several species of warbler, for example, can occupy the same acre of forest simultaneously by foraging for insects at different heights or on specific parts of the trees (figure 70).

Special Habitat Requirements

Some "ubiquitous" bird species are capable of living in a variety of forest types and successional stages. Other species, however, appear to be very sensitive to habitat changes and require specific habitat components. (See table 5, page 121.) Management goals for passerine birds can focus on particular "target" species or they can be generalized to accommodate a greater variety of species.

Management plans that target songbird populations must first consider the size of the managed parcel, and the nature of the surrounding properties. A 25 acre property surrounded by subdivision, for example, cannot realistically be managed to attract breeding pairs of worm-eating warblers or pileated woodpeckers. It can, however, attract flickers, hairy woodpeckers, scarlet tanagers, and many other species if the necessary habitat components are present. On the other hand, if the 25 acre parcel is adjacent to 300 additional wooded acres, it may well attract these more secretive, interior breeding birds.

Second, plans must consider the age, size, and species of forest stands, and the degree to which different-sized stands are interspersed (Capen 1982). Woodpeckers, in general, require snags and living trees with dead branches for food sources and nesting sites. Some woodpeckers have more specific habitat requirements than others. For example, pileated woodpeckers require mature forests with trees that host internal decay and colonies of carpenter ants, while flickers and red-headed woodpeckers prefer open woodlands

Management plans that target songbird populations must first consider the size of the managed parcel, and the nature of the surrounding properties.

interspersed with clearings. Abandoned cavities excavated by woodpeckers provide nesting and denning sites for numerous species of birds and mammals that cannot excavate their own.

Raptors

Hawks, owls, and other birds of prey are known as raptors. Approximately nineteen species breed in New England. (See table 6, page 122.) Several species migrate south before winter, and several arctic species migrate to the Northeast, for a winter total of approximately twenty-two species—depending upon the severity of winters and, specifically, upon the abundance and accessibility of food sources.

Most raptors have entirely predatory feeding habits. Major foods of raptors include birds, mammals, snakes, insects, fish, amphibians, and carrion. They usually occupy the highest trophic levels, although smaller raptors are sometimes preyed upon by larger raptors. Most owls are highly suited for hunting under low-light conditions, i.e., dawn, dusk, and night, while hawks are best suited for daytime feeding.

Most raptors have entirely predatory feeding habits.

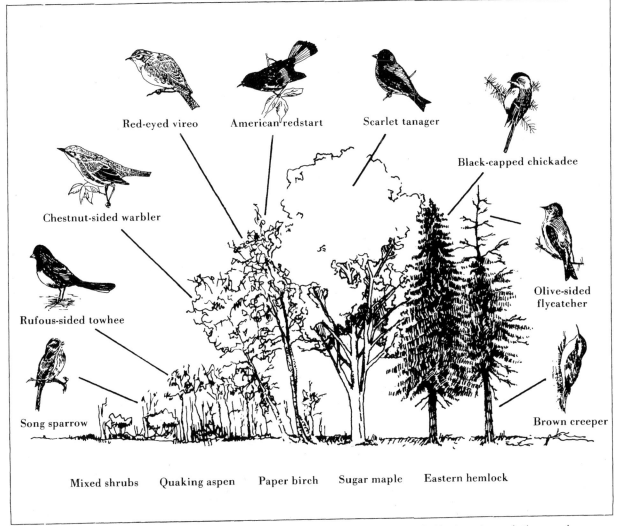

Figure 70. Each songbird species occupies a specific niche which separates its feeding habits from those of other species.

Chapter 5: Other Upland Forest Wildlife Species

Raptors are a necessary, important, and aesthetically pleasing component of forest ecosystems. Although they are patient hunters and often prey upon healthy individuals, they help dampen increases in prey populations and cull out physically disabled individuals.

Special Habitat Requirements

Cover and food resources adequate to sustain prey populations are necessary to attract and maintain raptors. Each species has specific habitat preferences in terms of successional stages, but open woodlands and clearings are preferred for hunting. Snags located in forest openings, along forest edges, and in woodlands with sparse understories provide ideal perching sites. Tall trees are preferred nesting sites for many raptors, and trees with large cavities are important nesting sites for kestrels (sparrow hawks) and owls (figure 71).

Managing Habitats for Upland Forest Wildlife

Managing for a high diversity of cover types in New England generally results in a high diversity of wildlife species. The elements of cover, food, and water (as described in Chapter 2) must be available to wildlife in sufficient quantity and quality in each habitat type. Management recommendations found in any wildlife plan will be a function of two things: the specific benefits of interest to the owner and the capacity of the land to produce those benefits. Wildlife biologists, state service foresters, or local consulting foresters can provide professional assistance in developing management plans that blend your interests with the potential of the land.

Management recommendations found in any wildlife plan will be a function of two things: the specific benefits of interest to the owner and the capacity of the land to produce those benefits.

Figure 71. Trees with large cavities are important nesting sites for owls.

Maintaining Early Successional Habitats for Food and Cover

Forests in the United States are sometimes interspersed with abandoned croplands, pastures, openings resulting from logging operations, and other early successional sites containing no trees or trees of seedling/sapling size. Such areas provide important habitats for many wildlife species and should be included in any wildlife management plan.

When the option exists, managing sites that are already in early stages of forest succession is much less costly and time-consuming than clearing land of trees to create forest openings. Also, openings that are maintained for woodcock, grouse, deer, and turkey can be managed for other wildlife species with little modification.

The method of maintaining early successional sites depends upon the desired objectives. Openings that are to be maintained strictly as herbaceous vegetation are managed differently from openings that are to be maintained as brush. Openings that contain only low herbaceous vegetation such as grasses are a vital habitat component, but they provide cover only at the ground level and lack mast (figure 72). Openings that contain both brushy cover and tall herbaceous growth, however, usually provide protective screening, palatable buds, berries, seeds, and a vertical dimension for additional nesting and perching opportunities (figure 73).

When the option exists, managing sites that are already in early stages of forest succession is much less costly and time-consuming than clearing land of trees to create forest openings.

Figure 72. Herbaceous openings contain abundant insect populations and are important seasonal habitat for many species.

Figure 73. Openings containing young woody brush provide additional, equally valuable food and cover.

Chapter 5: Other Upland Forest Wildlife Species

Maintaining Herbaceous Openings

Herbaceous openings (preferably smaller than 5 acres) are important for fulfilling seasonal habitat requirements for American woodcock, ruffed grouse, white-tailed deer, eastern wild turkeys, rabbits, hares, woodchucks, foxes, coyotes, songbirds, and raptors. They are also important as year-round habitats for small mammals such as meadow voles.

The most practical way to maintain herbaceous openings is by mowing every two to three years. Mowing should be delayed until after June when wild turkeys and most other birds have finished nesting. (First broods for songbirds producing more than one brood will also be out of the nest by this time.)

Herbaceous openings can be limed, fertilized, and seeded to increase their value as feeding sites. Before lime and fertilizer are applied, soils should be tested to ensure proper coverage. Lime and fertilizer, while not essential, do provide a temporary boost in producing lush, nutritious vegetation—an attractant for herbivores, insects, and the many species of birds and mammals that feed on them. Food plants desirable for seeding include clover, alfalfa, and the various seed mixes recommended by the USDA Soil Conservation Service. Seeding millet, sudan grass, sorghum, buckwheat, and other plants that retain seedheads into the autumn and above light snows can be effective for attracting wintering birds such as juncos, sparrows, finches, and cardinals. Seeds of native plants should be selected over exotic species whenever feasible.

Herbaceous openings can be limed, fertilized, and seeded to increase their value as feeding sites.

Figure 74. Five- to ten-year cutting rotations maintain brushy openings in various stages of development, increasing their habitat value.

Maintaining Brushy Openings

Forest openings containing a large percentage of shrubs and other woody vegetation attract a greater diversity of songbirds and fulfill more habitat requirements for more wildlife than permanent herbaceous openings (Probst 1979). They offer seasonal or year-round food and cover for woodcock, grouse, deer, wild turkeys, rabbits, hares, chipmunks, woodchucks, foxes, coyotes, songbirds, raptors, and small mammals such as shrews, meadow jumping mice, and white-footed mice.

Brushy openings (preferably less than 5 acres in size) can be managed by mowing or cutting on a rotational schedule of approximately five to ten years depending on the rate of growth. The openings can be divided into several convenient portions so that some portions are mowed/cut every other year, or into five portions so that one is mowed/cut annually. By this method, the openings eventually will be comprised of patches ranging from low herbaceous and woody ground cover to relatively tall brushy cover (figure 74).

Openings can be created in existing forests, but should not be created on steep slopes or on other sites that are likely to erode. Herbaceous openings should be created on sites where periodic mowing will be feasible. Openings should not expose ledges, caves, or streams; wide buffer strips should be maintained so that the quality of these features is not degraded for other wildlife or for the many plants that are unique to such sites. When feasible, cavity trees, snags, stubs, dense shrub patches, and fruit trees should be retained in the clearing and/or along its edges.

Sumacs, multiflora roses, ground junipers, eastern red cedars, barberries, raspberries, blackberries, some dogwoods, and some tree saplings and seedlings will readily appear in areas left unmowed, and provide excellent food and cover for wildlife.

Edges

The manner in which edges and their associated ecotones (see "Edges and Ecotones" in Chapter 2) are managed can significantly increase wildlife diversity. Edges having wide ecotones that gradually blend forest clearings with surrounding stands produce more cover and food for wildlife than edges having narrow, abrupt ecotones. Wide ecotones can be created around the periphery of clearings by leaving an unmowed strip at least 20 feet wide or as wide as is feasible (figure 75). Sprouts, saplings, shrubs, and tall herbaceous vegetation that grow in ecotones provide escape and resting cover for several wildlife species and provide feeding and nesting cover for edge species, including birds such as cardinals, catbirds, and song sparrows. Clumps of dense brush can be encouraged and retained as cover for rabbits and songbirds. Soft mast-producing shrubs, such as hawthorns, barberries, viburnums, honeysuckles, winterberries, and dogwoods should also be encouraged for their value as seasonal food sources.

Travel Lanes

In herbaceous clearings larger than 2 acres, or in uniform mature forests, brushy strips can be created sparingly to serve as travel lanes (see "Travel Lanes" in Chapter 2) connecting important habitat areas.

Edges having wide ecotones that gradually blend forest clearings with surrounding stands produce more cover and food for wildlife than edges having narrow, abrupt ecotones.

These linkage areas also provide some nesting, denning, escape, and resting cover for rabbits, hares, woodchucks, birds, and other wildlife. The width should be sufficient (15–60 feet) to provide adequate cover, and if within the forest, to allow sunlight to reach the ground. Travel lanes which meander somewhat are most valuable. Fencerows, stonewalls, and drainage ways surrounded by tall herbaceous vegetation and low woody growth make excellent travel lanes.

Cavity Nesters and Forest Openings

Cavity-nesting songbirds such as eastern bluebirds, tree swallows, and house wrens are easily attracted by the presence of cavity trees or nest boxes and snags around the periphery of clearings. Numerous nest box plans are available not only for songbirds, but for kestrels, screech owls, squirrels, bats, and others as well. State wildlife agencies and local nature centers may supply plans, materials, or even preconstructed nest boxes. Nest boxes should be cleaned annually by early March. To encourage second nesting attempts, bluebird boxes also should be cleaned after the first brood departs.

Managing Pole- and Timber-sized Forests for Food and Cover

Forests with high wildlife species diversity generally contain both well-developed understories and overstories (figure 76). Many of New England's forests are composed of uniformly aged stands as a

Cavity nesting songbirds such as eastern bluebirds, tree swallows, and house wrens are easily attracted by the presence of cavity trees or nest boxes and snags around the periphery of clearings.

Figure 75. Wide ecotones like the one in the background can be created by leaving an unmowed strip around a field's edge.

result of extensive clear-cutting and farm abandonment. The dense, uniform overstories of such stands can inhibit understory development and suppress wildlife species diversity and abundance.

Forests characterized by live trees 5–11 inches DBH are known as **pole stands**. They typically contain trees having straight, closely spaced trunks and minimal understories. **Timber-** or **sawtimber-sized forests** are characterized by live trees greater than 11 inches DBH and may lack well-developed understories as well. Late succession forests gradually develop shade-tolerant understories which replace the existing forest as old trees die and the canopy opens.

If wildlife species diversity is to be increased in intermediate and later succession forests, management plans should focus upon diversifying habitat conditions by encouraging multiple vegetative strata. Two different management strategies, even-aged management and uneven-aged management, can be used to accomplish similar results.

Even-aged Management

Forest stands dominated by trees of the same age are known as **even-aged stands**. Even-aged management is a cost-effective tool used for timber production, whereby entire stands are clear-cut as they reach maturity and are then replanted or regenerated. This method is well-suited to the oak and mixed hardwood forests which dominate much of southern and central New England. Forests are often managed so that several even-aged stands in different stages of development are present in one area.

If wildlife species diversity is to be increased in intermediate and later succession forests, management plans should focus upon diversifying habitat conditions by encouraging multiple vegetative strata.

Figure 76. Forests with high wildlife species diversity generally contain canopies of different heights within the same area.

Even-aged management can be a useful wildlife management tool regardless of whether timber production is planned. Even-aged management for wildlife can be accomplished by periodically clear-cutting 0.25–5 acre plots (patches or strips) in pole or timber stands and allowing them to pass through the different stages of forest succession. Because pole and timber stands often lack well-developed understories, they do not attract the wildlife species that require low canopies (less than 10 feet). By clear-cutting plots at annual intervals, however, various canopy heights and a greater diversity of habitat conditions become available for wildlife, especially for the many bird species preferring specific canopy heights (figure 77).

When clear-cutting new plots, some slash (tops and brush) should be retained on the site to provide cover and a source of invertebrates for mammals and birds. Brush can also be piled loosely to provide cover for mammals such as rabbits and for birds such as winter wrens. Brush piles should be placed at least 50 feet away from intended woodcock display sites, however. As clearcuts mature through sprout and sapling stages to pole size, the species of birds occupying the plots will change noticeably (DeGraaf 1982).

Uneven-aged Management

Forest stands that contain trees in a variety of age classes are known as **uneven-aged stands**. Uneven-aged management, as applied to timber production, encourages the growth of variously aged trees on the same site to ensure relatively short time frames between

As clearcuts mature through sprout and sapling stages to pole size, the species of birds occupying the plots will change noticeably.

Figure 77. These 1–2 acre patch cuts create early successional habitat in an otherwise uniformly mature forest.

harvests. Individual trees and/or small groups of trees are cut as they mature, releasing younger trees to grow for future timber production. Uneven-aged management can be practiced only with tree species which are at least moderately tolerant of shade.

Uneven-aged management can be suitable for wildlife management objectives because as mature trees are harvested, more sunlight and moisture reach the forest floor to promote understory development.

Overcrowded trees can be thinned and trees located where overstories are somewhat open can be felled sparingly to create slightly larger gaps. The desired result is to retain a large proportion of the overstory while simultaneously creating lower canopy layers to attract a wide variety of bird species (Probst 1979; Temple et al. 1979).

Unlike even-aged forests, uneven-aged forests contain canopies of various heights in the same stands. As long as these canopy layers remain relatively unchanged due to a balance of constant growth and periodic **selective cutting**, the same wildlife species will generally occupy the stands indefinitely. If the stands are not cut periodically, however, the overstory will grow dense; less sunlight and moisture will penetrate to the forest floor; the understory will thin and die; and the wildlife species will change significantly.

Cavity development can be encouraged by identifying and retaining (even in clearcuts) trees of high-cavity/low-timber potential.

Cavity Tree Management

Cavities in trees of all sizes are essential to many species of birds and mammals. For example, cavities having entrances less than 2 inches in diameter are important to nuthatches and white-footed mice, and well-developed cavities in large trees may be used by barred owls and gray foxes.* In some forests, particularly those having suffered heavy tree mortality from gypsy moths or similar pests, cavity trees will be ample. Timber management practices do not always favor the formation of cavities, however, because trees are harvested when they mature. Cavity development can be encouraged by identifying and retaining (even in clearcuts) trees of high-cavity/low-timber potential.

The decision as to which cavity trees to save should be based on the quality of the cavity and on the condition of the cavity tree. Live cavity trees usually stand longer than dead cavity trees and, therefore, provide long-term benefits (Gill 1982).

Snag and Stub Management

Dead, standing, leafless trees or broken trees with dead, standing trunks that are taller than 20 feet are known as **snags.** Broken trees having dead, standing trunks less than 20 feet tall are referred to as stubs. Snags and stubs are important foraging sites for many species of birds and often serve as cavity trees when primary excavators such as woodpeckers initiate cavity development. Snags—especially those with good vantage points in clearings or along edges—are also used as perching sites by raptors, phoebes, and other birds.

---

* *Living trees with cavities occupied by mammals are known as "den trees."*

Snags and stubs should be retained on all sites having a short supply. The number of snags required per given amount of acreage depends upon the desired wildlife species and on the landowner's objectives. (See table 7, page 123.) For example, one pair of downy woodpeckers requires at least four snags per 10 acres that are approximately 8 inches DBH and suitable for excavating (DeGraaf and Shigo 1985; Evans and Conner 1979). If high densities of downy woodpeckers are to be maintained, the number of snags per 10 acres must be increased to provide adequate foraging, roosting, and nesting sites while accounting for some snags that the woodpeckers find unsuitable. A recommendation to satisfy many snag-dependent species is to retain at least four snags per acre distributed according to other functional needs for safety, aesthetics, timber, and fuelwood production (Weber 1986).

In some forests, snags will be abundant naturally. Snags can be created by girdling or frilling poor-quality trees (see "Managing Winter Cover" for deer in Chapter 4). They can also be encouraged by extending the length of forest rotation periods. More wildlife will benefit from the creation of large snags (more than 18 inches DBH) as opposed to only numerous small ones. Large snags generally last longer and can be used by both large and small birds and mammals (DeGraaf and Shigo 1985; Evans and Conner 1979). Also, trees such as oaks with dense wood generally last longer as snags than trees such as aspen with softer wood.

Snags can be created by girdling or frilling poor-quality trees. They can also be encouraged by extending the length of forest rotation periods.

Conifers

Coniferous trees (those bearing needles and cones) comprise only one subgroup in the broad group of trees, shrubs, and vines labeled "evergreens." The term evergreen loosely refers to any species that retains its leaves or needles through winter. Common examples include mountain laurel, rhododendrons, American holly, partridgeberry, wintergreen, yew, and most conifers.

Conifers and other evergreens such as laurel are a necessary habitat component for many species of birds and mammals. They provide food and cover (dense foliage) for resting, roosting, nesting, and escaping, and especially help moderate the effects of inclement weather. Forests that contain both conifers and deciduous trees generally contain more wildlife species than forests containing either coniferous or deciduous trees exclusively (Capen 1979; Temple et al. 1979). Ruffed grouse, white-tailed deer, snowshoe hares, red squirrels, northern flying squirrels, martens, red-breasted nuthatches, golden- and ruby-crowned kinglets, solitary vireos, and blackburnian and bay-breasted warblers are examples of New England wildlife species attracted to conifers.

Small clumps of conifers, such as hemlock, white pine, and white spruce, typically can be planted in approximately 10-by-10 foot spacing and distributed at a ratio of approximately ½ acre per 10 acres of forest if they are lacking and are desired by the landowner. Encouraging natural regeneration of existing seed trees by releasing is often more practical and successful, however, especially in areas subject to deer browsing and gypsy moth defoliation. Conifers managed for ruffed grouse and white-tailed deer (as described in Chapters 3 and 4, respectively) also will suffice for other wildlife species.

Blowdowns

Trees that are uprooted and knocked down by strong winds provide an added source of cover in woodlands that lack substantial understories and dense brush (figure 78). By looking through the understories of snow-covered forests in winter, one can readily assess the amount of cover present. On most mature forest sites, the only cover dense enough to conceal deer, bobcats, grouse, hares, and other wildlife is young conifers, laurel, old tangles of vines and brush, and downed windblown trees (blowdowns). In many pole and mature stands, blowdowns can be the most prevalent source of winter cover.

We have a tendency to tidy our woods by cutting all "unsightly" blowdowns for firewood. By leaving some blowdowns (especially those which have retained their leaves) through their first winter, additional cover becomes available for birds and mammals—and the blowdowns can season for firewood.

To simulate blowdowns and create temporary sources of living cover, experienced chain saw operators can partially fell trees by cutting only part-way through the trunk. The fallen tree remains attached to the stump and will usually sprout vigorously and remain alive for several years. For best results, leaning trees should be selected over straight, upright trees, because they are less likely to snap off of their stumps and separate the **cambium**.

By leaving some blowdowns ... through their first winter, additional cover becomes available for birds and mammals—and the blowdowns can season for firewood.

Figure 78. Trees that are uprooted and/or blown over provide added cover which is particularly valuable in woodlands lacking a dense understory.

Trail Systems

A network of trails managed throughout one's property provides an ideal way to observe wildlife while walking, hiking, or cross-country skiing. Narrow, inconspicuous trails can easily be created by hand-pruning and require little maintenance. Larger trails can be created by clearing with a chain saw and can be maintained by mowing. Old logging roads, woodland access roads, unused town roads, and abandoned railroad beds are excellent segments to include in trail systems. Large areas of soil that become exposed during trail construction should be mulched and/or seeded with native grasses, legumes, or other forbs to minimize soil erosion and to provide an additional food source for wildlife.

Sections of trail that might suffer from erosion due to excessive drainage can be corrected with waterbars, terracing, diversions, and other means (*Timber Harvesting and Water Quality in Connecticut: A Practical Guide for Protecting Water Quality While Harvesting Forest Products*; Connecticut RC&D Forestry Committee, 1990). Waterbars can be created by digging shallow ditches across the trail at an angle or by partly burying a log approximately 10 inches in diameter across the trail at an angle (figure 79).

Trails should meander and pass adjacent to sites that are used most actively by wildlife. Meandering trails give the observer better chances for quietly approaching wildlife without being detected too soon. Some wide straightaways nearly 100 feet long that contain herbaceous ground cover offer good opportunities for spotting deer, rabbits, and foxes. Small openings less than ¼ acre can be created along trails to provide more feeding opportunities and an additional element of surprise for the observer and the observed.

Meandering trails give the observer better chances for quietly approaching wildlife without being detected too soon.

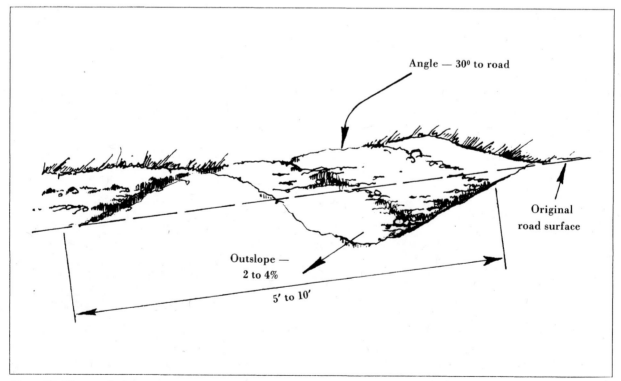

Figure 79. Where trails follow steep or lengthy slopes, diversions must be built to channel running water off the trail.

Table 5. Summary of forest habitat characteristics required by selected forest interior breeding birds.*

Species	Nest Location	Type	Feeding Location	Forest Size (Acres) Minimum	Optimum	Forest Age
Red-shouldered hawk	canopy	open	open areas	250	>250	mature
Barred owl	snag	cavity	open understory	250	>250	mature
Hairy woodpecker	snag	cavity	trunk	10	>40	mature
Pileated woodpecker	trunk	cavity	trunk	125	320	mature
Acadian flycatcher	shrub	open	subcanopy	80	95–300	mature
Yellow-throated vireo	canopy	open	canopy	250	10,000	mature
Red-eyed vireo	canopy	open	canopy	50	>250	mature
Northern parula warbler	ground	open	mid-story	750	>750	pole
American redstart	understory	open	mid-story	80	10,000	mature
Worm-eating warbler	ground	open	ground	750	2,500	mature
Swainson's warbler	understory	open	ground	—	800	mature
Ovenbird	ground	open	ground	250	6,500	mature
Kentucky warbler	understory	open	ground	80	>325	mature
Hooded warbler	understory	open	understory	80	1,500	mature
Scarlet tanager	canopy	open	canopy	25	250	mature

* Compiled from Bushman, E.S. and G. D. Therres, 1988. *Habitat Management Guidelines for Forest Interior Breeding Birds of Maryland.* Department of Natural Resources, Wildlife Tech. Pub. 88–1.

Chapter 5: Other Upland Forest Wildlife Species

Table 6. Raptors that breed in New England.*

Falcons	*Hawks*	*Osprey*	*Owls*	*Vultures*
American kestrel	**Accipiters**	Osprey	Eastern screech owl	Turkey vulture
Merlin	Sharp-shinned hawk		Great horned owl	
Peregrine falcon	Cooper's hawk		Barred owl	
	Northern goshawk		Long-eared owl	
			Northern saw-whet Owl	
	Buteos			
	Red-shouldered hawk			
	Broad-winged hawk			
	Red-tailed hawk			
	Eagles			
	Bald eagle			

* Compiled from DeGraaf, R. and D. Rudis, 1986. *New England Wildlife: Habitat, Natural History, and Distribution.* USDA Forest Service GTR NE–108.

Table 7. Number of cavity trees[a] needed to sustain the hypothetical maximum populations of nine species of woodpeckers found in New England.

Species	Territory size (Acres)	Average nest tree[b] DBH (Inches)	Average nest tree[b] Height (Feet)	(A) Cavity trees used, minimum (Number)	(B) Pairs/ 100 acres, maximum (Number)	Cavity trees needed/ 100 acres[c] (A x B)
Red-headed woodpecker	10	20	40	2	10	20
Red-bellied woodpecker	15	18	40	4	6.3	25
Yellow-bellied sapsucker	10	12	30	1	10	10
Downy woodpecker	10	8	20	4	10	40
Hairy woodpecker	20	12	30	4	5	20
Three-toed woodpecker	75	14	30	4	1.3	5
Black-backed woodpecker	75	15	30	4	1.3	5
Northern flicker	40	15	30	2	2.5	5
Pileated woodpecker	175	22	60	4	0.6	2.4

a. After Evans and Conner (1979).
b. Larger trees may be substituted for smaller trees.
c. Number of cavity trees needed to sustain population at hypothetical maximum level.

Review Questions

1. Why are herbaceous openings important to upland forest wildlife? How can openings be maintained?

2. Why are seedling/sapling-sized brushy areas important to upland forest wildlife? How can such areas be maintained?

3. What is a travel lane?

4. What distinguishes even-aged from uneven-aged management? Under what conditions can uneven-aged management be practiced?

5. Why are cavity trees important for upland forest wildlife?

6. What are "snags"? Why are they important?

Field Exercises

1. Obtain a photocopy of the forest cover type map that you sketched for chapters 1 and 2. Try to estimate the percentage of total land area covered by herbaceous openings (fields, woods, roads, old log landings, etc.).

    If the total area is less than 10%, have your service or consulting forester help you identify areas that could be converted to herbaceous openings, i.e., mowable areas with little timber or other value.

2. Visit each of the existing herbaceous openings and examine their edges. Classify the ecotones (transition areas between cover types) as wide and gradual or as narrow and abrupt.

    What species typify these areas?

    What might you do to widen abrupt ecotones?

3. From your map, estimate the percentage of total area for which the forest cover is:
    1. seedling/sapling-sized (dominant trees less than 5 inches DBH)
    2. pole-sized (dominant trees 5–11 inches DBH)
    3. timber-sized (dominant trees over 11 inches DBH)

    Which of these cover types is in short supply? Discuss with your forester how you can manage your forest long-term to better balance these size classes.

These exercises were designed to be done on your own woodland. Once completed, you will be well on your way toward your own wildlife management plan.

Field Notes:

6 Wetlands Wildlife

Numerous wildlife species are restricted to wetlands habitats and depend upon the sustained welfare of such habitats.

Introduction

The wetlands that dot forests in the Northeast are widespread and often highly productive habitats that deserve special recognition in forest management plans for wildlife. Numerous wildlife species are restricted to wetland habitats and depend upon the sustained welfare of such habitats.

Wetlands historically have been looked upon as sites to fill or ignore. The total area once occupied by wetlands in the United States has been estimated at 127,000,000 acres. By 1950, an estimated 45,000,000 acres, or 35%, had been drained (Shaw and Fredine 1956). Ongoing losses have been estimated at least 300,000 acres annually (Weller 1987).

Chapter 6 presents small-scale management practices designed to enhance wetlands for wildlife and to buffer them against the activities on adjacent lands. The basic types of wetlands in the northeastern United States are introduced, and wildlife species associated with particular wetland types are discussed.

What Are Wetlands?

Definition of Wetlands

Wetlands are areas that become saturated with water for all or part of the growing season. By a U.S. Fish & Wildlife Service definition, "Wetlands are transitional lands between terrestrial and aquatic systems where the water table is usually at or near the surface or the land is covered by shallow water" (Cowardin et al. 1979). The definition of wetlands varies slightly by region. One example is the classification of wetlands in Connecticut that is based on soil characteristics. The Connecticut Inland Wetlands and Watercourses Act (1972) defines wetlands as "land . . . which consists of any of the soil types designated as poorly drained, very poorly drained, alluvial, and flood plain by the National Cooperative Soil Survey, as may be amended from time to time, of the U.S. Soil Conservation Service of the United States Department of Agriculture" (Ammann et al. 1986).

Description of Wetland Types

This chapter focuses only on freshwater wetlands, not on coastal or tidal wetlands which contain saltwater or **brackish** waters. The "wetlands" addressed in this chapter include inland wetlands and deepwater habitats (Cowardin et al. 1979).

Wetlands are areas that become saturated with water for all or part of the growing season.

Figure 80. Example of a stunted red maple swamp.

Inland Wetlands

Inland wetlands include lands that are seasonally flooded or that maintain year-round water levels shallow enough to support **emergent vegetation** during the growing season. Emergent vegetation has submerged root systems but some foliage and especially flower parts that "emerge," or grow above the water surface. Specific site conditions such as water depth, degree of acidity or alkalinity, soil type, nutrient availability, oxygen level, and exposure to sunlight determine which plant species are capable of inhabiting each wetland. Based on all of these conditions plus the plant species present, one can determine the wetland type. Systems for classifying inland wetlands can be somewhat complicated, so this chapter abbreviates the classification systems and includes three broad types—swamps, marshes, and bogs.

Inland wetlands dominated by woody vegetation are known as swamps.

Swamps

Inland wetlands dominated by woody vegetation are known as swamps. Shrub swamps are inland wetlands dominated by woody vegetation less than 20 feet tall (Cowardin et al. 1979). Due to restrictions imposed by certain site conditions, the vegetation in some shrub swamps remains less than 20 feet tall indefinitely. Other relatively new shrub swamps are temporary and produce trees greater than 20 feet tall as forest succession proceeds. Stunted red maple swamps are perhaps the most common examples of shrub swamps in New England, especially in southern regions (figure 80). Other woody species

Figure 81. Forested wetlands with sparse overstory and lush undergrowth.

found in shrub swamps include speckled alder, willow, elder, red-osier dogwood, eastern white cedar, northern white cedar, highbush blueberry, swamp rose, sweet pepperbush, and poison sumac.

Forested Wetlands

Swamps dominated by trees taller than 20 feet are known as forested wetlands (Cowardin et al. 1979). Such swamps are often subject to significant fluctuations in water level and may periodically lack standing (surface) water. Typical tree species include red maple, swamp white oak, black gum, black ash, green ash, eastern white cedar, northern white cedar, and black spruce. Lush understories comprised of species such as sweet pepperbush, winterberry, poison sumac, and spicebush are often present in forested wetlands having sparse overstories (figure 81). Woody understories are frequently absent in forested wetlands having dense overstories, but herbaceous plants as skunk cabbage, jack-in-the-pulpit, and cinnamon fern are often present.

Vernal and Ephemeral Pools

In forested wetlands possessing water tables located at or just below the ground's surface, standing water may be visible only for a few weeks after snowmelt or immediately following heavy rains. Such standing waters are known as vernal (springtime) and ephemeral pools, respectively (figure 82). They are mentioned separately in this chapter so that their importance to wildlife is not overlooked. Many

In forested wetlands possessing water tables located at or just below the ground's surface, standing water may be visible only for a few weeks after snowmelt or immediately following heavy rains.

Figure 82. Vernal pool in woodland with light snowcover.

frogs and salamanders require such temporary habitats specifically for breeding and egg-laying. They also provide drinking sites for many birds and mammals and often provide unfrozen sites around which early spring arrivals of woodcock can probe for food.

Marshes

Freshwater marshes are inland wetlands characterized by a predominance of herbaceous emergent vegetation. The division of marsh types as follows is based on the widely used classification system of Martin et al. (1953) as re-published by Shaw and Fredine (1956).

Deep Marshes

Deep marshes almost always contain standing water and sometimes contain pockets of **open water**, or water with no emergent or floating vegetation. Typical plants include emergents such as cattails and bulrushes, and floating species such as the fragrant white water lily and the yellow pond lily.

Shallow Marshes

Shallow marshes contain less standing water than deep marshes and, therefore, are more likely to dry up during droughts. They frequently border deep marshes (figure 83). Common plants include pickerelweed, arrow arum, arrowhead, bur reed, cattails, wetland grasses, bulrushes, spike rushes, and blue flag.

Shallow marshes contain less standing water than deep marshes and, therefore, are more likely to dry up during droughts.

Figure 83. Shallow marsh with dense vegetation (background) bordering deep marsh.

Wet Meadows

Wet meadows are marshes having water tables at or just below the ground's surface. They may be saturated with water only seasonally and may contain standing water following heavy precipitation or stream overflow. Wet meadows typically support water-loving species of grasses, sedges, and rushes, as well as numerous showy wildflowers (figure 84).

Bogs

Bogs are inland wetlands characterized by acidic and nutrient-deficient waters. Such characteristics tend to exclude the presence of specialized bacteria and other organisms necessary for the decomposition of organic matter. Decomposition occurs at very slow rates, and dead organic matter accumulates as "peat" instead of being recycled as available nutrients for future use by plants and animals. Plant diversity in bogs is usually low compared to other inland wetlands, and the plants that are able to withstand such acid and nutrient-poor environments are highly specialized and often restricted to bog ecosystems. Typical plant species found in bogs include sphagnum mosses, cranberry, leatherleaf, sheep laurel, pitcher plants, sundews, cotton grass, rose pogonia, black spruce, and tamarack.

The amount of peat and live vegetation in bogs varies greatly. Some bogs are firm and safe to walk on (figure 85) while other bogs possess either thin floating mats of entwined roots or open waters with advancing mats. Bogs become firm as more and more peat accumulates beneath the mat and as the root layer becomes thicker. Caution should always be exercised when walking on bogs so that observers do not break through the mat and so that minimal damage occurs to the bog's unique plant life.

Plant diversity in bogs is usually low compared to other inland wetlands.

Figure 84. Blue flag, a common resident of wet meadows.

Figure 85. Northern New England spruce bog.

Chapter 6: Wetlands Wildlife

Deep-Water Habitats

Deep-water habitats include permanent open waters that are too deep (usually deeper than 6.6 feet) to support emergent vegetation (Cowardin et al. 1979). Common examples of deep-water habitats include lakes, reservoirs, ponds, and rivers. Swamps, marshes, and bogs are often present along shores and islands of deep-water habitats (figure 86).

Sunlight barely penetrates water past a depth of 6.6 feet. Below this depth, waters tend to be relatively dark and poorly oxygenated. Some plants, such as algaes and duckweeds, cope with such environmental limitations by floating freely on or near the water's surface where light and oxygen are most abundant. Other plants anchor onto the **substrate** and grow upward in the water until their foliage reaches sufficient light. In comparison to the fewer and broader leaves of many terrestrial herbaceous plants, such species often have numerous thin leaves which serve to increase the plants' surface area for more efficient oxygen intake. Typical aquatic plants found in deep-water habitats include algaes, duckweeds, pondweeds, bladderworts, and milfoil.

Deep-water habitats include permanent open waters that are too deep . . . to support emergent vegetation.

Figure 86. Deep-water habitat bordered by deep and shallow marshes.

132

Significance of Inland Wetlands to Wildlife

The transition between terrestrial and aquatic communities usually produces an abundance of cover, food, and water—the habitat elements essential to wildlife. The rich habitats of inland wetlands attract insects, amphibians, reptiles, fish, birds, and mammals which are all part of a giant food web in the wetland ecosystem. Some species, such as wood ducks, black ducks, mallards, green-backed herons, least bitterns, Virginia rails, marsh wrens, sedge wrens, river otters, muskrats, and beaver, live almost entirely in wetland habitats. Other species, such as common yellowthroats, red-winged blackbirds, red-shouldered hawks, marsh hawks, white-tailed deer, bobcats, mink, and raccoons, frequent wetlands for cover, food, or water.

The Role of Inland Wetlands

Most inland wetlands contain waters that have been trapped by low or level topography and unforgiving bedrock. As a result, they serve as storage basins and dampen the effects of excess **runoff** and floodwaters. Wetlands also serve as sediment traps; erosion from upslope or upstream contributes substantially to the rate at which wetlands become filled in by sediments. Some wetlands also may play important roles in recharging **aquifers** from which we obtain well water.

Most inland wetlands contain waters that have been trapped by low or level topography and unforgiving bedrock.

Selected Species

Wood Duck

Wood ducks nearly became extinct in the early twentieth century due to excessive unregulated hunting, the draining of lowlands for agriculture, and the cutting of lowland timber. The passing of the Migratory Bird Law of 1913 and the Migratory Bird Treaty Act of 1918 coupled with the construction of artificial nest sites enabled wood ducks to recover dramatically. Today the wood duck is one of the most common ducks in the ***Atlantic flyway*** (figure 87). They are present during the breeding season throughout the Northeast, excluding northern Maine, and begin migrating to southeastern portions of the United States by October.

Adult wood ducks feed primarily on aquatic and terrestrial plant matter throughout the year. Acorns are a primary autumn food source, and hickory nuts, grapes, mulberries, and other hard and soft mast are eaten when available. Young wood ducks initially feed on aquatic and terrestrial invertebrates and gradually increase their intake of plant foods (Bellrose 1976). Wood ducks nest in tree cavities and artificial nest boxes, sometimes several hundred feet from open water.

Special Habitat Requirements

Wood ducks rely on marshes, swamps, slow-moving streams, and ponds with shallow water to provide emergent vegetation and aquatic invertebrates on which to feed. Mast-producing trees and shrubs bordering wetland habitats are important for supplementing summer and autumn diets. Cavity trees exceeding 16 inches DBH with nest boxes or cavity entrances at least 4 inches in diameter are crucial for nesting.

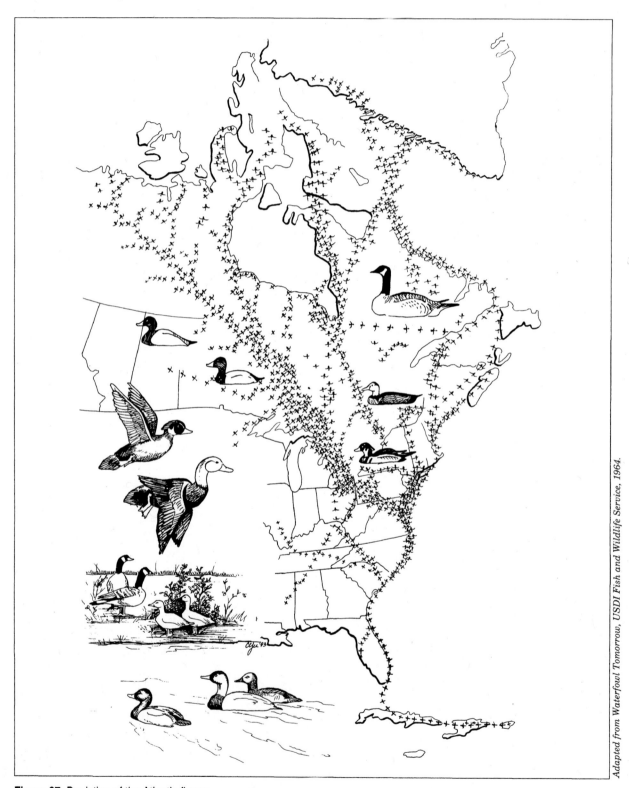

Figure 87. Depiction of the Atlantic flyway.

Other Cavity-nesting Waterfowl

Besides the wood duck, three other cavity-nesting species of *waterfowl* breed in the Northeast—the common goldeneye, hooded merganser, and common merganser. Two additional cavity nesters, the bufflehead and Barrow's goldeneye, winter in New England. The common goldeneye and common merganser breed in northern New England, and the hooded merganser breeds in all but extreme southern New England (Peterson 1980; DeGraaf and Rudis 1986).

In fresh water, common goldeneyes feed on snails, crayfish, insect larvae, and other aquatic invertebrates and on the roots, stems, leaves, and seeds of aquatic plants. Both mergansers feed on fish, crayfish, frogs, and aquatic insect larvae. All three species generally breed on fresh water and winter on unfrozen salt water, rivers, lakes, and reservoirs.

Special Habitat Requirements

The common goldeneye, hooded merganser, and common merganser breed on lakes, ponds, and rivers. For nesting, common goldeneyes generally show a preference for cavity trees at least 20 inches DBH located near or over water. Hooded mergansers have nest site requirements similar to wood ducks and may similarly select sites located considerable distances from water. Common mergansers usually nest in cavity trees greater than 20 inches DBH but will resort to ground nesting if cavity trees are scarce (Bellrose 1976; Pough 1953). Each species, particularly hooded mergansers, will nest in artificial nest boxes.

American Black Duck

The American black duck population has declined in recent years due to a combination of factors. Habitat loss, the eastward expansion of the mallard (and subsequent competition for nest sites where habitats overlap), and cross breeding between the two species are hypothesized as major causes (Heusmann 1982). The black duck breeds throughout New England and winters in southern and coastal areas (figure 88).

Black ducks consume a wide variety of foods. Fresh-water diets include a great diversity of aquatic invertebrates and the stems, leaves, and seeds of aquatic plants. Diets occasionally include amphibians, fish, and mast such as acorns and tupelo seeds. Manure and waste grain spread on agricultural fields may provide winter food when other foods are in short supply. Black ducks usually nest on the ground within the concealment of vegetation but may infrequently nest on stumps and in large cavities, especially in flooded timber (Cowardin et al. 1967).

Black ducks and wood ducks commonly feed in water by tipping their bodies forward so that all but their tails and hind third of their bodies are submerged. They exemplify a feeding behavior characteristic of the group of ducks known as "dabblers," which are also known as "marsh ducks" or "puddle ducks." Unlike the goldeneyes and mergansers, which belong to the group of waterfowl known as "divers," dabblers can spring directly into flight from water. The legs

Black ducks usually nest on the ground within the concealment of vegetation but may infrequently nest on stumps and in large cavities, especially in flooded timber.

of divers are positioned far back on the body to aid in underwater swimming, and most divers need to run on the water before becoming airborne.

Special Habitat Requirements

Black ducks nest in isolated swamps, in fresh and brackish marshes, in bogs, and along swampy or marshy borders of deep-water habitats. Weedy and somewhat brushy borders along ponds generally enhance the suitability of sites for nesting. Waters shallow enough to support aquatic vegetation are important for providing foods such as duckweed and coontail and the seeds of pondweeds, yellow pond lily, sedges, and smartweeds (Bellrose 1976).

Other Ground-nesting Waterfowl

Approximately nine additional species of ground-nesting waterfowl breed regularly in the Northeast: the mute swan; Canada goose; dabblers (i.e., mallard, green-winged teal, blue-winged teal, northern shoveler, and gadwall); and divers (i.e., ring-necked duck and red-

> *Waters shallow enough to support aquatic vegetation are important for providing foods such as duckweed and coontail and the seeds of pondweeds, yellow pond lily, sedges, and smartweeds.*

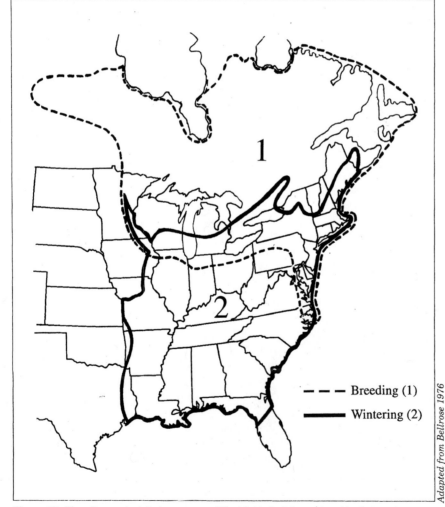

Figure 88. Breeding and wintering ranges of the black duck in eastern North America.

breasted merganser). The mallard breeds throughout New England. The mute swan, Canada goose, blue-winged teal, northern shoveler, and gadwall breed in southern and/or coastal New England. The green-winged teal, ring-necked duck, red-breasted merganser, plus the blue-winged teal breed in northern New England (Peterson 1980; DeGraaf and Rudis 1986). Some of these species winter south of New England, and many western and northern species migrate to New England for winter.

The foods and feeding habits of the mute swan, Canada goose, and dabblers are similar to those of the wood duck and American black duck. Canada geese often "graze" on lawns to obtain tender shoots, and northern shovelers commonly feed on **plankton** by straining water through their broad, elongated bills. Foods and feeding habits of breeding ring-necked ducks and red-breasted mergansers are similar to those of common goldeneyes and common mergansers, respectively.

Special Habitat Requirements

Swamps, marshes, open-water bogs, deep-water habitats bordered by swampy or marshy shorelines, and the coves of slow-moving rivers fulfill the habitat requirements of many ground-nesting waterfowl species during the breeding season. Sites that receive little disturbance are generally preferred. To satisfy the feeding requirements of most species, sites should contain sections of shallow water which produce an abundance of aquatic vegetation and invertebrates. Sufficient low, woody, or herbaceous cover must be present along shorelines and islands to provide nesting opportunities. Shorelines that are devoid of vegetation due to heavy grazing or mowing are generally unsuitable for nesting, as are shorelines comprised solely of dense tangles of woody growth which might exclude a vegetative transition from forest to water.

Herons and Bitterns

The herons and bitterns live exclusively in wetland habitats. Five herons (the great blue, little blue, green-backed, black-crowned night heron, and yellow-crowned night heron) and two bitterns (the American and least) regularly inhabit inland wetlands of the Northeast. The great blue heron and American bittern are present throughout New England; the green-backed heron, black-crowned night heron, and least bittern are distributed throughout southern, central, and coastal New England; and the little blue heron and yellow-crowned night heron occur sporadically along coastal and southern New England. Herons and bitterns are migratory, although some great blues and black-crowned night herons winter along coastal southern New England.

Herons and bitterns are predatory. Major foods in their diverse diet include fish, crayfish, frogs, salamanders, birds, small mammals, insects, spiders, and leeches (DeGraaf and Rudis 1986). Herons generally construct nests in trees in the vicinity of open water (figure 89) and often form nesting colonies known as heronries. Bitterns tend to construct well-hidden ground nests in dense herbaceous growth near open water.

Swamps, marshes, open-water bogs, deep-water habitats bordered by swampy or marshy shorelines, and the coves of slow-moving rivers fulfill the habitat requirements of many ground-nesting waterfowl species during the breeding season.

Special Habitat Requirements

Herons utilize a variety of wetland habitats provided that feeding sites remain productive. Shallow waters must be present to permit wading and feeding, and sufficient emergent or aquatic vegetation must also be present to provide habitats for prey. Bitterns are secretive and generally prefer the dense cover provided by tall herbaceous vegetation associated with marsh environments.

Rails and Grebes

The secretive rails and grebes live almost exclusively in wetland habitats. Three species of rails—the Virginia rail, king rail, and sora—breed in inland marshes (figure 90). The pied-billed grebe is the only species of grebe that breeds in New England. The Virginia rail and sora inhabit all of New England, and the king rail occurs in the southern half of New England. The pied-billed grebe breeds throughout New England, excluding northern Maine.

Rails feed on aquatic and terrestrial insects, snails, and other aquatic invertebrates; small fish; seeds from plants such as sedges; and occasionally on waste grain from nearby agricultural fields (DeGraaf and Rudis 1986). They tend to be unobtrusive in all of their activities and prefer to run rather than fly when disturbed. Unlike rails, which usually feed by wading and probing, grebes tend to feed by diving and swimming. Primary foods include small fish, tadpoles,

Shallow waters must be present to permit [herons'] wading and feeding, and sufficient emergent or aquatic vegetation must also be present to provide habitats for prey.

Figure 89. Two young great blue herons (left) stand by their nest in the rookery. Nearby an adult great blue heron sits on a nest in the crotch of trees (right).

aquatic insects, snails, leeches, and other aquatic invertebrates. Grebes can remain underwater for long periods of time if disturbed and often surface in the middle of a dense clump of emergent vegetation for cover. Rails construct ground nests that are well concealed under dense vegetation, while grebes build mound-shaped ground nests that are anchored to emergent vegetation.

Special Habitat Requirements

Rails and grebes inhabit marshes and marshy borders of deep-water habitats. During migration, rails often stop at wet meadows, and grebes often rest and feed on open waters of deep-water habitats. An abundance of marsh vegetation growing in shallow water of consistent depth is important to both, especially during the breeding season.

During migration, rails often stop at wet meadows, and grebes often rest and feed on open waters of deep-water habitats.

Muskrat

The muskrat is a medium-sized, semiaquatic rodent with a sparsely haired tail. Muskrats are distributed throughout the Northeast and may be common to abundant in favorable habitats.

They feed primarily on vegetation such as cattails, bur reed, bulrushes, arrowhead, pickerelweed, duckweed, pondweed, and water lilies and on succulent shoots and fruits of terrestrial plants (Godin 1977). Insects, crayfish, clams, snails, mussels, frogs, sluggish fish, reptiles, birds, and carrion are consumed to a lesser extent (Godin 1977).

Figure 90. Some New England marsh birds: (clockwise from top left) king rail, Virginia rail, pied-billed grebe, and sora.

Muskrats excavate underwater tunnels and dens in banks or construct lodges from emergent vegetation when suitable banks are not adjacent to wetlands. Lodges are built in shallow water and appear as mounds up to 9 feet in diameter made of stems and roots from cattails and other emergents (figure 91). Smaller mounds are used as shelters in which to eat while protected from predators and inclement weather. Muskrats remain active in **runs** under ice, feeding on stored roots and stems or on vegetation near their den or lodge. They live in bank dens and lodges through the winter, and if water levels fluctuate, may construct new lodges on top of frozen wetlands by gnawing through the ice and mounding vegetation (Errington 1961; Godin 1977). Muskrats assist in maintaining open waters for waterfowl, but they can burrow into dams and eat excessive amounts of emergent vegetation if they become overpopulated.

Special Habitat Requirements

Muskrats inhabit swamps, marshes, bogs, and shorelines of deepwater habitats with consistent water levels. Shallow waters having ample quantities of emergent and/or aquatic vegetation must be present.

Muskrats excavate underwater tunnels and dens in banks or construct lodges from emergent vegetation when suitable banks are not adjacent to wetlands.

Figure 91. Muskrat lodge in marsh. Note herbaceous plant materials used.

Photo: J. S. Barclay

Beaver

Beaver were nearly exterminated from the United States by unregulated fur harvests in the 1800s but have recovered remarkably due to trapping restrictions and reintroductions. They are the largest North American rodent, possessing broad, flattened, nearly hairless tails. Adult beaver in the Northeast typically weigh from 27 to 67 pounds (Godin 1977), but adults may attain weights in excess of 100 pounds (Wade and Ramsey 1986). Beaver are distributed throughout New England.

Favored foods include the bark of deciduous trees and, to a lesser extent, of conifers. Herbaceous plants such as bulrushes, sedges, coontails, and pond lily roots are also consumed in spring and summer (Godin 1977). Beaver excavate dens in banks or construct lodges depending upon the site characteristics of the wetland. Lodges appear as mounds of logs, branches, and mud approximately 25 feet in diameter (figure 92). Like the muskrat, beaver develop networks of underwater runs and live in bank dens or lodges through the winter, feeding on the bark of logs and branches that have been **cached** in or along runs, tunnels, and lodges. The dams that beaver construct not only regulate water flow and depth for ideal beaver habitat but also create habitats for many other wetland species of wildlife and serve as effective flood-control structures.

Special Habitat Requirements

Beaver utilize a variety of inland wetlands and relatively shallow deep-water habitats for dam and lodge construction. Sufficient densities of favored trees and shrubs must surround potential sites. Aspens, poplars, willows, alders, and birches less than fifteen years old seem to be preferred.

The dams that beaver construct not only regulate water flow and depth for ideal beaver habitat but also create habitats for many other wetland species of wildlife and serve as effective flood-control structures.

Figure 92. Beaver lodges appear as mounds of logs, branches, and mud.

Aquatic Mustelids

Mink and river otter belong to the family Mustelidae, which also includes the upland mustelids—that is, weasels, marten, fisher, and striped skunk. Both species are distributed throughout New England.

Both are predatory and feed on a wide variety of prey species. Typical foods include fish, frogs, crayfish, clams, snails, salamanders, turtles, snakes, earthworms, and insects. Mink also prey on small mammals, muskrats, rabbits, and birds and frequently cache their food in crevices, hollow stumps, or other protected sites (Godin 1977). Mink and river otter are adept swimmers.

Special Habitat Requirements

Mink and river otter inhabit various types of inland wetlands and edges of deep-water habitats. They may travel several miles across uplands to reach new habitats. Waters must be productive enough to sustain an abundance of prey. Hollow logs; hollow stumps; spaces under large roots; fallen trees; rock crevices; and abandoned beaver, muskrat, or woodchuck burrows near water are important for denning, although they can excavate their own dens.

Raccoons

Raccoons are frequent visitors and inhabitants of forested inland wetlands. They are common in New England and are distributed throughout.

Raccoons are omnivorous and opportunistic feeders. Typical foods include hard and soft mast, buds, tender shoots, agricultural crops, crayfish, frogs, snails, terrestrial and aquatic insects, snakes, turtles, small mammals, birds, eggs, and carrion. Raccoons are adept tree climbers. They generally remain in a state of dormancy from late November to late March and survive on stored fat (Godin 1977).

Special Habitat Requirements

The presence of productive inland wetlands and deep-water habitats is generally sufficient to attract raccoons. Large cavities in trees exceeding approximately 20 inches DBH and located near water are important for denning. Caves, man-made structures, woodchuck burrows, and beaver and muskrat lodges serve as alternative denning sites.

Managing Forest Wetland Habitats for Wildlife

Inland wetlands and deep-water habitats can be managed as separate units within forest management plans. Management activities in adjacent forests should be conducted in such a manner that wetlands are not adversely affected.

It is important to become familiar with the appropriate state inland wetlands regulations before implementing management plans in wetland habitats. Regulations differ between states, and permits

It is important to become familiar with the appropriate state inland wetlands regulations before implementing management plans in wetland habitats.

may be required before habitats can be modified in any major way. Connecticut, for example, has a law prohibiting the clear-cutting of timber on inland wetlands without a permit.

This section is designed to acquaint the private landowner with relatively inexpensive, small-scale management practices used to enhance wetlands for wildlife. Potentially costly, large-scale projects such as pond and dam construction are not discussed. For those interested in large-scale projects, pertinent references are listed under "Literature Cited & Selected References."

Maintaining Buffer Strips and Corridors

One of the most important management practices for forest inland wetlands and deepwater habitats is protection from activities on surrounding land. Improper woodland access road construction uphill and adjacent to wetlands can lead to soil erosion and increase sedimentation in waters. Adjacent tree harvesting on steep slopes can also result in soil erosion and can allow more nutrients and sunlight to reach wetlands, causing increased water temperatures and algal blooms. Grazing in and on the banks of wetlands can cause soil erosion and eliminate vegetation essential to wetland wildlife.

Buffer strips are effective solutions for isolating wetlands from detrimental activities. Retaining an uncut or unmowed strip at least 50 feet wide surrounding wetlands provides a minimal but crucial degree of protection. Buffer strips serve as filters for reducing erosion and sedimentation, provide shade trees to help maintain cooler water temperatures, provide visual screening, and, combined with fencing, provide an adequate barrier from livestock.

Buffer strips serve as filters for reducing erosion and sedimentation, provide shade trees to help maintain cooler water temperatures, provide visual screening, and, combined with fencing, provide an adequate barrier from livestock.

Figure 93. Careful planning of haul road and/or skid trail layout will minimize stream and wetland impacts from logging.

Buffer strips serve as important travel lanes and sources of cover and food for many wetland and upland species of wildlife. Streamside buffer strips are especially important as ***riparian corridors***, i.e., travel lanes that allow wildlife access from one habitat to another (Harris 1985; Craven et al. 1987). Corridors can be established by maintaining buffer strips through fields, clearcuts, or backyards and by connecting all buffer strips to create a continuous network. Corridors are particularly important to mammals with large home ranges such as river otter, mink, white-tailed deer, bobcat, and lynx and to reptiles, amphibians, and migrating birds.

Measures to Prevent Erosion and Sedimentation

Soil erosion and sedimentation in wetlands can be minimized by proper location and construction of forest roads, skid trails, and clearings. Roads that must parallel inland wetlands and streams should be located on the forested side of buffer strips, never within the buffer strip or in place of a buffer strip. The recommended width of buffer strips for road construction varies with the site and, according to one set of guidelines (Miller et al. 1979), depends on soil erosion potential and the slope of surrounding land (Hynson et al. 1982). (See table 8, page 149.) The greater the potential for soil erosion and the steeper the slope, the wider the buffer strips must be for adequate pro-tection.

When planning a skid-trail network in forests, it is important that roads and trails cross streams in as few places as possible (Hassinger et al. 1981). A map depicting proposed roads in relation to inland wetlands, streams, and other habitat components is invaluable in making planning decisions (figure 93). If roads must cross streams, bridges and culverts should be installed after a hay bale barrier or similar sediment control structure is in place, to avoid muddying waters during each crossing. Hynson et al. (1982), the Connecticut RC&D Forestry Committee (1990), and the Vermont Agency of Environmental Protection (undated) provide options for stream crossings and illustrations showing proper construction. Crossings should be perpendicular to the stream flow and located on a straight, narrow portion of the stream on low banks comprised of firm rocky soil (Hynson et al. 1982). Roads can be seeded with native grasses and legumes to minimize erosion and to provide food sources for grouse, deer, turkey, rabbits, and other wildlife. Freshly exposed soil should be mulched quickly under most circumstances.

Managing Cover and Food Plants

In ideal marsh situations, the transition between forested buffer strips and wetlands proper is gradual, ranging from trees in the buffer strip, to a shrub zone, to a tall emergent vegetation zone, to a shorter herbaceous zone, and finally to open water (figure 94). Most shrub swamps, forested wetlands, and bogs lack such transition.

For the more terrestrial species of wildlife that visit wetlands only on an occasional or irregular basis, buffer strips serve as important sources of protective cover. The attractiveness of buffer strips to many wildlife species can be enhanced by encouraging herbaceous food

Soil erosion and sedimentation in wetlands can be minimized by proper location and construction of forest roads and clearings.

plants and mast-producing trees and shrubs. Soft-mast producing shrubs such as winterberry, dogwoods, blueberries, viburnums, and elder can be released as necessary to stimulate mast production and vegetative growth. Tall, overhanging trees that border swamps and marshes and suppress the growth of shrubs and herbaceous vegetation can be removed occasionally to increase sunlight. Younger trees in shrub swamps can be removed around open pockets of water to retard succession and develop a wider herbaceous border as a transition from water to trees.

In marshes, tall emergents such as cattails can be cut during periods of low water in winter or drought to create temporary patches of open water or to increase the area of existing open water, particularly for waterfowl. Growth will be inhibited during the following growing season if water levels rise, and the water may remain open for several years (Weller 1978; Weller 1987). Floating vegetation such as pond lilies can be cut low with a scythe to create open water on a short-term basis, usually for one growing season or less in shallow waters.

Planting desired wetland vegetation to increase cover and food can be successful but is often difficult if plants are already established in the target planting site (Yoakum et al. 1980). Planting should be conducted only as a last resort to vegetation encouragement. Species to be planted should be specifically matched to the geographic location and to the exact characteristics of the site as determined by professional assistance. Arrowhead and duckweed can be introduced by transplanting entire plants in early spring; pondweed and smartweed

Planting desired wetland vegetation to increase cover and food can be successful but is often difficult if plants are already established in the target planting site.

Figure 94. Transitional vegetation from open water (center left) to marsh (foreground) and forest (background).

Chapter 6: Wetlands Wildlife

can be planted from rootstocks in early spring; and sedges and grasses can be planted from rootstocks or entire plants in spring or early summer (Yoakum et al. 1980). Small, 0.1–1.0 acre plots adjacent to wetlands (on sites with low erosion potential) can be tilled and seeded with millets, buckwheats, corn, or other foods to attract songbirds, grouse, and turkey and to experiment with the response by waterfowl. Many waterfowl species are readily attracted to large agricultural fields but are sometimes reluctant to feed in small food plots.

Managing for Cavity Trees, Snags, and Beaver

Cavity trees and snags are important habitat components of many wetlands. Mature trees with large cavities around the periphery of wetlands are likely to house raccoons or barred owls. Snags in and around wetlands are ideal perching sites for raptors, kingfishers, and songbirds and are likely to attract woodpeckers, the primary excavators.

Cavity trees and snags should be encouraged and retained in buffer strips. Managing for a diversity of cavity sizes will benefit a broader range of wildlife species. Cavities with entrance holes up to 2 inches in diameter may be used by white-footed mice, swallows, eastern bluebirds, and chickadees; cavities with intermediate-sized entrance holes may be used by cavity-nesting waterfowl, smaller owls, and flying squirrels. Similarly, different-sized snags often attract different species of woodpeckers which excavate holes of distinct sizes.

Cavity trees and snags should be encouraged and retained in buffer strips. Managing for a diversity of cavity sizes will benefit a broader range of wildlife species.

Figure 95. Just-hatched downy duckling in wood duck nestbox opening.

Photo: U.S. Department of Agriculture, Soil Conservation Service

In wetland habitats there probably is no such thing as too many cavity trees or snags. Trade-offs may exist, however. For example, managing for wood ducks and other cavity-nesting waterfowl could result in the retention of too many large cavity trees used by raccoons, since raccoons are a major nest predator of waterfowl. The landowner's objectives must be carefully weighed. Poor-quality trees can be girdled or frilled to create snags of a particular size if more are desired.

Beaver can be nuisances if they establish residence where they are unwanted, but the habitats that they create often become highly productive. Flooded timber that is not felled by beaver for dam or lodge construction eventually dies, and the growth of herbaceous vegetation usually increases on banks and silt deposits in response to increased sunlight. More cavities become available for wildlife as woodpeckers excavate nest holes in the newly-created snags and then abandon them. Beaver impoundments increase habitat diversity and provide new habitats for wetland wildlife, "terrestrial" cavity nesters, and fish.

Artificial Nest Structures for Inland Wetlands

Cavity-nesting waterfowl, particularly wood ducks and hooded mergansers, can be attracted to artificial nest boxes attached to trees or posts installed near the edges of open water (figure 95). Boxes can be placed on metal poles or wood posts, but an inverted metal cone should be attached as a predator guard if wood posts are used. Three or four inches of coarse sawdust or wood shavings should line each box, and should be changed annually. Nest success tends to be higher when two to four boxes per acre are installed (Bellrose 1976). Smaller nest boxes can be installed to attract smaller cavity nesters; even eastern bluebirds are common in certain wetlands.

In wetland habitats there probably is no such thing as too many cavity trees or snags. Trade-offs may exist, however.

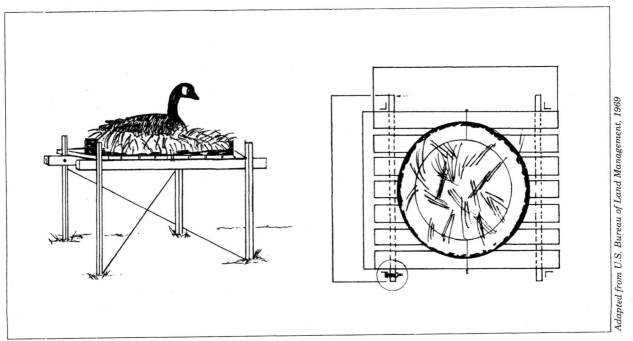

Figure 96. Nesting platform for Canada goose, showing construction detail.

Adapted from U.S. Bureau of Land Management, 1969

147

Raised platforms work well as nesting sites for Canada geese and mallards. Platforms can be made of wood and mounted on posts tall enough to raise the platform 1–3 feet above land or water (figure 96). A tire can then be wired to the top of the platform and covered with hay or straw to provide the start of a nest (Yoakum et al. 1980). The platform can be installed at the edge of a wetland or several feet offshore on an island in emergent vegetation.

Trail Systems

Carefully planned trail systems offer a rewarding way to observe wetland wildlife. Narrow trails can be created in buffer strips and can extend to the edge of wetlands at locations most likely to afford good views of a wood duck's favorite nesting site or of a great blue heron's most productive hunting spot. Vistas can be created by clearing enough brush so that observers can look out but are not readily detected by the observed. If a trail is to receive frequent use, foot bridges and wooden walkways can be constructed to minimize erosion and other disturbances over streams and spongy areas.

Carefully planned trail systems offer a rewarding way to observe wetland wildlife.

A Note on Pond Design

If ponds are to be created, several guidelines for the benefit of wildlife can be considered. Lawns mowed to pond edges may attract geese but otherwise appear to be of little value to wildlife. Buffer strips of natural vegetation along the pond edge attract wildlife and can add aesthetic value as well. Ponds with irregular, meandering configurations provide more edge for aquatic and terrestrial plants and animals. Islands contribute additional edge and can provide nesting sites for waterfowl with some degree of protection from predators. Water depth is crucial for plant growth and, therefore, to the animals that depend on plants for cover and food. Shallow sections of ponds are necessary for the presence of emergents; deep sections are important to fish and wildlife during periods of low water levels.

Local, state, and in some cases federal permits may be required for pond construction in an increasing number of states, depending upon the type of activity and its effects on existing wetlands or watercourses. Be sure to consult with local officials before starting work.

Table 8. An example of recommended buffer strip widths (in feet*)[1].

Erosion Hazard of Soil	Percent Slope						
	0	10	20	30	40	50	60
Slight	30	55	80	105	130	155	180
Moderate	40	75	100	140	170	200	235
Severe	50	90	130	170	210	250	290

* Double these distances for disturbed areas in municipal water supply waterbeds.

1. Guidelines for Virginia (Hynson et. al 1982) (after Miller et. al 1979).

Review Questions

1. What is the difference between inland wetlands and deep-water habitats?

2. How does the vegetation in a swamp differ from the vegetation in a marsh?

3. What are the basic habitat requirements of "dabblers"?

4. How, and in what ways, do buffer strips protect wetlands?

5. What are some precautions that can be used to minimize soil erosion adjacent to wetlands?

Field Exercises

1. With a copy of the forest cover type map that you sketched for Chapters 1 and 2, visit your property to determine if your original wetland delineations are accurate. Sketch in and redraw areas that you might not have included as wetlands, e.g., forested wetlands and vernal pools.

2. Visit any wetland that might be present on your property and begin to assemble a list of the plants and wildlife in it, using field guides and binoculars if necessary.

    Based on the characteristics of the wetland's edge and on the plants observed, what wildlife species would you expect to find in your wetland?

    How many species of wildlife do you see directly or indirectly—remember that tracks, scats, and feeding evidence are important signs of presence.

    Do you notice any cavity trees, snags, bank dens, lodges, or runs?

3. Obtain a photocopy of the forest cover type map that you sketched for Chapters 1 and 2 and also your **Woodcock Habitat Map**, **Grouse Habitat Map**, and **Deer and Turkey Habitat Map**.

    If you sketched access roads on your forest cover type map, do you feel that they are properly located based on what you have learned? Are they situated correctly in respect to wetlands, slope, and intensive management areas?

    Sketch new roads on your forest cover type map if you feel that existing roads are improperly located.

    For those who do not have roads, sketch what you think might be good locations for some.

    Follow the same procedures for trails. Do the trails include views of wetlands and management areas for woodcock, grouse, deer, and turkey? Do they give access to features of interest to you?

    Talk with your service or consulting forester about constructing any new access roads or trails that you wish to install.

These exercises were designed to be done on your own woodland. Once completed, you will be well on your way toward your own wildlife management plan.

Field Notes:

Appendix

Table 9. List of scientific names of plants mentioned in the text.

Name Used in Text	Common Name of Intended Species	Scientific Name
Alder	common alder	*Alnus serrulata*
	speckled alder	*Alnus rugosa*
Alfalfa	alfalfa	*Medicago sativa*
Apple	apple	*Malus spp.*
	wild apple	*Malus pumila*
	flowering crab	*Malus coronaria*
Arrow arum		*Peltandra virginica*
Arrowhead		*Sagittaria latifolia*
Ash	white ash	*Fraxinus americana*
	black ash	*Fraxinus nigra*
	green ash	*Fraxinus pennsylvanica*
Aspen	bigtooth aspen	*Populus grandidentata*
	quaking aspen	*Populus tremuloides*
Barberry	European barberry	*Berberis vulgaris*
	Japanese barberry	*Berberis thunbergii*
Beech	American beech	*Fagus grandifolia*
Birch	black birch	*Betula lenta*
	gray birch	*Betula populifolia*
	white birch	*Betula papyrifera*
	yellow birch	*Betula alleghaniensis*
Bittersweet	American bittersweet	*Celastrus scandens*
Bladderwort	swollen bladderwort	*Utricularia inflata*
Blackberry	blackberry	*Rubus spp.*
Black cherry	black cherry	*Prunus serotina*
Black gum (Tupelo)		*Nyssa sylvatica*
Blueberry	highbush blueberry	*Vaccinium corymbosum*
	lowbush blueberry	*Vaccinium angustifolium*
Blue flag		*Iris versicolor*
Buckwheat		*Fagopyrum spp.*
Bulrush		*Scirpus spp.*
Bur reed		*Sparganium americanum*
Cattail	broadleaf cattail	*Typha latifolia*
	narrowleaf cattail	*Typha anqustifolia*
Cedar	Atlantic white cedar	*Chamaecyparis thyroides*
	eastern red cedar	*Juniperus virginiana*
	northern white cedar	*Thuja occidentalis*
Cherry	black cherry	*Prunus serotina*
	choke cherry	*Prunus virginiana*
	pin cherry	*Prunus pennsylvanica*
Cinnamon fern		*Osmunda cinnamomea*
Clover	red clover	*Trifolium pratense*
	white clover	*Trifolium repens*
Common spatterdock		*Nuphar advena*
Coontail		*Ceratophyllum spp.*

Table 9. List of scientific names of plants mentioned in the text (continued).

Name Used in Text	Common Name of Intended Species	Scientific Name
Corn	corn	*Zea mays*
Cotton grass		*Eriophorum polystachion*
Cranberry		*Vaccinium macrocarpon*
Dogwood	alternate-leaved dogwood	*Cornus alternifolia*
	bunchberry	*Cornus canadensis*
	flowering dogwood	*Cornus florida*
	gray-stemmed dogwood	*Cornus racemosa*
	narrowleaf dogwood	*Cornus obliqua*
	red-osier	*Cornus stolonifera*
	round-leaved dogwood	*Cornus rugosa*
	silky dogwood	*Cornus amomum*
Duckweed		*Lemna spp.*
Elder or Elderberry	common elder	*Sambucus canadensis*
	red-berried elder	*Sambucus pubens*
Elm	American elm	*Ulmus americana*
	slippery elm	*Ulmus rubra*
Fir	balsam fir	*Abies balsamea*
Grape	grape	*Vitis spp.*
Greenbrier	greenbrier	*Smilax spp.*
Hawthorn	hawthorn	*Crataegus spp.*
Hazel	hazelnut	*Corylus americana*
	beaked hazelnut	*Corylus cornuta*
Hemlock	eastern hemlock	*Tsuga canadensis*
Hickory	bitternut hickory	*Carya cordiformis*
	mockernut hickory	*Carya tomentosa*
	pignut hickory	*Carya glabra*
	shagbark hickory	*Carya ovata*
Honeysuckle	fly honeysuckle	*Lonicera canadensis*
	tatarian honeysuckle (escape)	*Lonicera tatarica*
Hop-hornbeam	American hop-hornbeam	*Ostrya virginiana*
Hornbeam	American hornbeam	*Carpinus caroliniana*
Jack-in-the-pulpit		*Arisaema triphyllum*
Jewelweed	pale jewelweed	*Impatiens pallida*
	spotted jewelweed	*Impatiens capensis*
Juniper	common juniper	*Juniperus communis*
Laurel	mountain laurel	*Kalmia latifolia*
	sheep laurel	*Kalmia angustifolia*
Leatherleaf		*Chamaedaphne calyculata*
Lily	fragrant water lily	*Nymphaea odorata*
	small pond lily	*Nuphar microphyllum*
	yellow pond lily	*Nuphar variegatum*
Locust	black locust	*Robinia pseudoacacia*
Maple	red maple	*Acer rubrum*
	sugar maple	*Acer saccharum*
Meadowsweet	meadowsweet	*Spiraea latifolia*
Milkweed	common milkweed	*Asclepias syriaca*
Millet	barnyard grass	*Echinochloa crusgalli*
	Japanese millet	*E. crusgalli var. frumentacea*

Appendix

Table 9. List of scientific names of plants mentioned in the text (continued).

Name Used in Text	Common Name of Intended Species	Scientific Name
Oak	black oak	*Quercus velutina*
	northern red oak	*Quercus rubra*
	scarlet oak	*Quercus coccinea*
	white oak	*Quercus alba*
Panicum grass	panicum	*Panicum spp.*
Pea	wild pea (and others)	*Lathyrus spp*
Pickerelweed		*Pontederia cordata.*
Pine	jack pine	*Pinus banksiana*
	pitch pine	*Pinus rigida*
	eastern white pine	*Pinus strobus*
	red pine	*Pinus resinosa*
Pitcher plant		*Sarracenia purpurea*
Poison ivy	poison ivy	*Toxicodemdron radicans*
Raspberry	raspberry	*Rubus spp.*
Rhododendron	rhododendron	*Rhododendron spp.*
Rose	rose	*Rosa spp.*
	swamp rose	*Rosa palustris*
Rose pogonia		*Pogonia ophioglossoides*
Rye	common rye	*Secale cereale*
Sedge	sedge	*Carex spp.*
Sensitive fern	sensitive fern	*Onoclea sensibilis*
Shadbush	shadbush	*Amelanchier spp.*
Skunk cabbage		*Symplocarpus foetidus*
Sphagnum		*Sphagnum spp.*
Spicebush		*Lindera benzoin*
Spikerush		*Eleocharis spp.*
Spruce	black spruce	*Picea mariana*
	red spruce	*Picea rubens*
	white spruce	*Picea glauca*
Strawberry	strawberry	*Fragaria spp.*
Sumac	poison sumac	*Toxicodendron vernix*
	smooth sumac	*Rhus glabra*
	staghorn sumac	*Rhus typhina*
Sundew	round-leaved sundew	*Drosera rotundifolia*
	thread-leaved sundew	*Drosera filiformis*
Sweet fern	sweet-fern	*Comptonia peregrina*
Sweet pepperbush		*Clethra alnifolia*
Tamarack		*Larix laricina*
Tupelo	black gum	*Nyssa sylvatica*
Viburnum	viburnum	*Viburnum spp.*
	black-haw	*Viburnum prunifolium*
	downy arrow-wood	*Viburnum rafinesquianum*
	highbush-cranberry	*Viburnum trilobum*
	hobblebush	*Viburnum alnifolium*
	maple-leafed viburnum	*Viburnum acerifolium*
	nannyberry	*Viburnum lentago*
	northern arrow-wood	*Viburnum recognitum*
	possum haw	*Viburnum nudum*
Violet	violet	*Viola spp.*

Table 9. List of scientific names of plants mentioned in the text (continued).

Name Used in Text	Common Name of Intended Species	Scientific Name
Virginia creeper	Virginia creeper	*Parthenocisus quinquefolia*
Walnut	black walnut	*Juglans nigra*
	northern wild raisin	*Viburnum cassinoides*
Wheat	wheat	*Triticum spp.*
Willow	willow	*Salix spp.*
	beaked willow	*Salix bebbiana*
	black willow	*Salix nigra*
	pussy willow	*Salix discolor*
	shining willow	*Salix lucida*
Winterberry	winterberry	*Ilex verticillata*
Witch hazel	witch hazel	*Hamamelis virginiana*

Appendix

Table 10. List of scientific names of animals mentioned in the text.

Name Used in Text	Common Name of Intended Species	Scientific Name
American woodcock	American woodcock	*Scolopax minor*
Arctic tern	arctic tern	*Sterna paradisaea*
Bat	big brown bat	*Eptesticus fuscus*
	eastern pipistrelle	*Pipisstrellus subflavus*
	Indiana myotis	*Myotis sodalis*
	Keen's myotis	*Myotis keenii*
	little brown myotis	*Myotis lucifugus*
	small-footed myotis	*Myotis leibii*
Bear	black bear	*Ursus americanus*
Beaver		*Castor canadensis*
Bittern	American bittern	*Botaurus lentiginosus*
	least bittern	*Ixobrychus exilis*
Bobcat	bobcat	*Felis rufus*
Brown thrasher	brown thrasher	*Toxostoma rufum*
Bufflehead		*Bucephala albeola*
Canada goose		*Branta canadensis*
Cat	bobcat	*Lynx rufus*
	domestic cat	*Felis domestica*
Chickadee	black-capped chickadee	*Parus atricapillus*
Chipmunk	eastern chipmunk	*Tamias striatus*
Cliff swallow	cliff swallow	*Etrochelidon pyrrhonota*
Common snipe	common snipe	*Capella gallinago*
Common yellowthroat		*Geothlypis trichas*
Coyote	coyote	*Canis latrans*
Crow	crow	*Corvus brachyrhynchos*
Curlew	eskimo curlew	*Numenius borealis*
	long-billed curlew	*Numenius americanus*
Deer	white-tailed deer	*Odocoileus virginianus*
Deer mouse	deer mouse	*Peromyscus maniculatus*
Dowitcher	long-billed dowitcher	*Limnodromus scolopaceus*
	short-billed dowitcher	*Limnodromus griseus*
Duck	American black duck	*Anas rubripes*
	ring-necked duck	*Aythya collaris*
	wood duck	*Aix sponsa*
Eagle	bald eagle	*Haliaeetus leucocephalus*
	golden eagle	*Aguila chrysaetos*
Eastern bluebird		*Sialia sialis*
Fisher		*Martes pennanti*
Fox	gray fox	*Urocyon cinereoargenteus*
	red fox	*Vulpes vulpes*
Gadwall		*Anas strepera*
Godwit	Hudsonian godwit	*Limosa haemastica*
	marbled godwit	*Limosa fedoa*
Goldeneye	Barrow's goldeneye	*Bucephala islandica*
	common goldeneye	*Bucephala clangula*
Golden plover	lesser golden plover	*Pluvialis dominica*
Grouse	ruffed grouse	*Bonasa umbellus*
Gypsy moth	gypsy moth	*Lymantria dispar*

Table 10. List of scientific names of animals mentioned in the text (continued).

Name Used in Text	Common Name of Intended Species	Scientific Name
Hawk	Cooper's hawk	*Accipiter cooperii*
	marsh hawk	*Circus cyaneus*
	northern goshawk	*Accipiter gentilis*
	red-shouldered hawk	*Buteo lineatus*
	sharp-shinned hawk	*Accipiter striatus*
Heron	black-crowned night heron	*Nycticorax nycticorax*
	great blue heron	*Ardea herodias*
	green-backed heron	*Butorides striatus*
	little blue heron	*Egretta caerulea*
	yellow-crowned night heron	*Nycticorax violaceuss*
Hummingbird	ruby-throated hummingbird	*Archilochus colubris*
Lynx		*Felix lynx*
Mallard	mallard (duck)	*Anas platyrhynchos*
Marten		*Martes americana*
Merganser	common merganser	*Mergus merganser*
	hooded merganser	*Lophodytes cucullatus*
	red-breasted merganser	*Mergus serrator*
Mink	mink	*Mustela vison*
Muskrat		*Ondatra zibethicus*
Mute swan		*Cygnus olor*
Northern oriole	"Baltimore" oriole	*Icterus galbula*
Northern shoveler		*Anas clypeata*
Opossum	Virginia opossum	*Didelphis virginiana*
Owl	barred owl	*Strix varia*
	great horned owl	*Bubo virginianus*
	screech owl	*Otus asio*
Pied-billed grebe		*Podilymbus podiceps*
Porcupine	porcupine	*Erethizon dorsatum*
Rabbit	eastern cottontail	*Sylvilagus floridanus*
	New England cottontail	*Sylvilagus transitionalis*
	snowshoe hare	*Lepus americanus*
Raccoon	raccoon	*Procyon lotor*
Rail	king rail	*Rallus elegans*
	Virginia rail	*Rallus limicola*
Raven	northern raven	*Corvus corax*
Red-winged blackbird		*Agelaius phoeniceus*
River otter		*Lutra canadensis*
Robin	American robin	*Turdus migratorius*
Sandpiper	sandpiper	*Calidris spp.*
		Actitis spp.
		Tringa spp.
		Micropalama spp.
		Tryngites spp.
		Bartramia spp.
Skunk	striped skunk	*Mephitis mephitis*
Snake	northern black racer	*Coluber c. constrictor*
	black rat snake	*Elaphe o. obsoleta*
Spruce budworm	spruce budworm (moth)	*Choristoneura fumiferana*

Appendix

Table 10. List of scientific names of animals mentioned in the text (continued).

Name Used in Text	Common Name of Intended Species	Scientific Name
Squirrel	gray squirrel	*Sciurus carolinensis*
	red squirrel	*Tamiascuirus hudsonicus*
Starling	European starling	*Sturnus vulgaris*
Sora		*Porzana carolina*
Teal	blue-winged teal	*Anas discors*
	green-winged teal	*Anas crecca*
Turkey	wild turkey	*Meleagris gallopavo*
	eastern wild turkey	*Meleagris gallopavo*
Turkey vulture	turkey vulture	*Cathartes aura*
Weasel	long-tailed weasel	*Mustela frenata*
	short-tailed weasel	*Mustela erminea*
Wolf	gray or "timber" wolf	*Canis lupus*
Woodcock	American woodcock	*Scolopax minor*
Wood frog	wood frog	*Rana sylvatica*
Wren	marsh wren	*Cistothorus palustris*
	sedge wren	*Cistothorus platensis*
Yellow-bellied sapsucker		*Sphyrapicus varius*

Table 11. Metric conversions.

From U.S. Customary System units		To SI (metric) units		Multiply by
Unit	Abbreviation	Unit	Abbreviation	U.S. to SI Conversion Factor
Area				
acre	A	hectare	ha	0.4047
square foot	ft^2	square meter	m^2	0.0929
square inch	in^2	square centimeter	cm^2	6.4516
square mile	mile^2	square kilometer	km^2	2.5900
Length				
foot	ft	meter	m	0.3048
inch	in	centimeter	cm	2.54
micron	micron	micrometer	μm	1
mile	mile	kilometer	km	1.6093
yard	yd	meter	m	0.9144
Mass				
ounce	oz	gram	g	28.3495
pound	lb	kilogram	kg	0.4536
Temperature				
degrees Fahrenheit	°F	degrees Celcius (or Centigrade)	°C	$t_{°C} = (t_{°F} - 32) \div 1.8$
Volume				
bushel	bu	liter	L	35.2391
gallon	gal	liter	L	3.7854
ounce	oz	milliliter	mL	29.5735
quart	qt	liter	L	0.9464

Glossary

A
Age class. A group of individuals that are of a similar age in a population.
Aquifers. Natural ground water systems.
Atlantic flyway. One of the major migratory routes of waterfowl. Other flyways include the Mississippi flyway and Pacific flyway.

B
Bed. To rest; describes the action of deer when resting or sleeping, usually in a location that provides protection, high outward visibility, and easy escape routes.
Brackish. Mixing with salt water. Brackish waters typically have salinities higher than fresh waters but lower than salt waters.
Browse. Palatable twigs, shoots, and buds of woody plants.
Browse line. In forest stands, an obvious horizontal separation between healthy vegetation and vegetation devoid of edible twigs, shoots, and buds as a result of heavy feeding pressure by deer, usually in winter.
Brush hogging. Clearing or maintaining early successional sites with a rotary mower (known as a "brush hog") pulled behind a tractor.
Buffer strip. A protective border of trees, brush, and/or herbaceous vegetation along streams and other sensitive sites.

C
Cache. *Verb:* to hide food for future use. *Noun:* a food supply hidden and stored for future use.
Cambium. A layer of living cells located between the inner bark and wood of trees. Cambium is responsible for outward growth.
Carnivores. Organisms that feed primarily on animal flesh.
Carrying capacity. The number of individuals (in a population) that an ecosystem can support over time.
Caruncles. Fleshy, knobby protuberances on the necks of male turkeys.
Catkins. Cylindrical, scaly, seed heads (flower heads) of certain deciduous trees, e.g., birches, willows, aspens, and alders.
Cavities. Hollows in living or dead trees that usually form following injury to the tree or excavation by woodpeckers. Cavities provide important nesting and resting cover for many birds and mammals.
Clear-cutting. In a wildlife management sense, removing all trees and brush in a forested area for the purpose of creating early successional sites. Sawlogs and fuel wood are valuable by-products.
Clearings. Forest-land sites containing no, or few, trees.
Climax forest. A stand of trees dominated by species that represent the final stage in forest succession for a given site.
Clone. A group of trees originating from root sprouts produced by a single tree.
Community. The living portion of an ecosystem, often referred to in terms of several populations in a specific area.
Compensatory mortality. The death of "surplus" individuals in a population as determined by the limits of carrying capacity.
Conifers. Trees having cones and needles or scale-like leaves which, in most species, are

retained throughout the winter (examples include spruce, fir, pine, cedar, juniper, and larch). The wood of conifers is referred to as "softwood."

Corridors. Wide travel lanes that allow access from one site to another, as from a bedding site to a feeding site.

Cover. Plants or other structures that are used by animals for shelter and protection.

Cover types/forest types. Forest-land classifications based on vegetative composition.

D

Deciduous. Referring to trees that shed their leaves annually (as opposed to conifers, which, in general, are "evergreen").

Dewlap. Loose, unfeathered skin that hangs from the throats of male turkeys.

Dispersal. The movement of an animal from one area (generally the vicinity of the place of birth or hatching) to another for the purpose of exploiting new habitat resources (when establishing new territories, for example).

Diversity. The number of species and individuals of various species present in an area.

Drumming log. Any site selected by a ruffed grouse as a stage for drumming and displaying. The site might be a log, stump, stonewall, or boulder.

E

Ecosystem. A complex of interactions between living and nonliving components of the environment.

Ecotone. The area directly effected by edges; the area of transition between one plant community and an adjacent plant community of different species composition or age.

Edges. Borders located where two different plant communities or similar communities of different ages join.

Emergent vegetation. Plants having roots that grow beneath water but foliage that grows above water.

Even-aged stands. Forest stands dominated by trees of the same age class.

Exclosures. Fences designed to keep deer out of an area (as opposed to *enclosures,* which keep animals in a specific area).

F

Flights. Loose groups of migrating woodcock.

Food chain. An orderly progression of successive foods eaten within a community, usually starting with plants and "ending" with a carnivore.

Food web. The combination of all food chains within a community.

Forbs. Palatable, broad-leaved, herbaceous vegetation.

Forest succession. The natural progression and replacement of plant species over time as determined by environmental conditions and by the plants themselves.

G

Girdling (frilling). Creating a continuous ring around a tree with a chain saw, axe, or hatchet to sever the cambium and kill the tree.

Gobbling. Vocalization of male turkeys advertising their presence to females and other males, especially during the breeding season. From a distance, gobbling can sound similar to the incessant barking of a small dog.

Guard object. A structure such as a tree, stump, or boulder that probably provides a sense of security to incubating ground-nesting birds and drumming grouse.

H

Habitat. Any area that contains all resources essential to the survival of a wildlife population.

Hardwoods. Deciduous trees.

Glossary

Herbaceous plants. Non-woody plants.

Herbivores. Animals that feed primarily on plants.

Herds. Deer populations in a particular area—town, country, or state, for example.

Hibernacula. Caves used by bats for winter hibernation. Large numbers of bats often congregate at favorite hibernacula.

Home range. The area occupied by an animal when it performs its daily functions and travels to cover, food, and water.

I

Inbreeding. The interbreeding of related individuals of the same species; often results in undesirable genetic traits or abnormalities.

Intermediate species. Trees that tolerate some degree of full sun and full shade. Intermediate species typically dominate forest development between pioneer and climax stages of succession.

Intermittent streams. Water courses that flow only at certain times of the year, i.e., during periods of snowmelt and high water tables.

Interspersion. The degree to which various cover types are present in an area—in contrast to a habitat consisting of one uniform cover type, such as a continuous stand of oak and hickory. The amount of edge present is indicative of the degree of interspersion.

Invertebrates. Species in the animal kingdom which do not possess a backbone. In the context of this publication, the term invertebrate refers to insects, spiders, centipedes, millipedes, and earthworms.

K

Keratin. Durable, fibrous proteins.

L

Leaf litter. The accumulation of leaves, twigs, and other organic matter on the forest floor.

Limiting factor. Any habitat element (cover, food, or water) that is in short supply and thereby hinders a species' ability to survive in a given area.

Locally. Spottily; present only in areas supporting the necessary combination of habitat requirements.

Logging headers (landings). Central clearings in a logging operation that are convenient for skidding, collecting, loading, and hauling saw logs.

Longevity. The expected life span of an individual.

M

Mast. The seed and fruit of a tree or shrub. Hard mast includes nuts; and soft mast includes catkins, berries, and other fruits.

Molting. In birds, the process of feather replacement. Old, worn feathers loosen and drop as new feathers replace them; or brilliantly colored feathers may simply replace unworn, drab-colored feathers in preparation for the breeding season. Most birds molt completely or partially once or twice every year.

N

Nematodes (roundworms). Cylindrical, unsegmented worms. Only some species digest injured wood and assist in the formation of cavities.

Niche. The role of a species in a community.

O

Omnivores. Animals that feed readily on both plants and other animals.

Open water. Portions of water that remain free of emergent or floating vegetation.

Opportunistic feeders. Animals that consume a wide range of plant and animal foods as they are found or as they become available.

Overstory. The highest canopy, or layer of tree crowns, in a forest.

Overtopping trees. Trees growing above shorter or younger trees. The crowns of overtopping trees form an overstory which hinders sunlight penetration to the understory.

P

Persistent fruits. Berries and other fruits that are retained on the plant long after ripening.

Photosynthesis. The process by which green plants utilize the sun's energy to combine carbon dioxide and water into simple sugars for food.

Pioneer species. Plants that require full sunlight and initially grow on bare soils following major disturbances such as fires and bulldozing. Pioneer species quickly die in the shade of intermediate species.

Plankton. Minute plants and animals found in water.

Pole stands. Forest stands dominated by trees 5–11 inches in diameter 4½ feet above ground.

Precocial. Referring to chicks that are well developed at hatching and capable of leaving the nest within minutes, as opposed to altricial chicks (songbird chicks, for example), which are helpless at hatching and remain in the nest for several days.

Predator. An animal that captures and consumes other animals.

Prey. An animal that is captured by a predator for consumption.

Primary feathers. The outermost flight feathers of a bird's wing when spread.

R

Regenerate. To encourage a new tree crop by removing older trees. Regeneration occurs from stump sprouts (coppice) or from seed germination.

Release cutting (releasing). Removal of competing trees, shrubs, or vines to promote the growth of a desired plant species.

Reservoirs. Bodies of water (impoundments) created by man-made water-control structures such as dams.

Riparian corridors. In strictly a wildlife management sense, buffer strips that serve as travel lanes along streams and rivers. For a particular species, corridors allow access from one habitat to another across non-habitat areas.

Riparian zones. Flood plains, banks, and associated areas that border free-flowing or standing water.

Roosting fields. Early successional forests, pastures, unimproved hay fields, abandoned agricultural fields, or other fields used by woodcock for nocturnal roosting.

Rotations. Intervals of time between successive clearcuttings on the same site. Rotation length is determined by the age at which a stand of trees reaches optimum condition for a desired goal.

Rubs. Portions of saplings debarked or abraded by male deer rubbing their antlers against them.

Runoff. For the purpose of this publication, precipitation that does not enter the soil but flows down slope, possibly causing erosion and carrying sediments into streams and inland wetlands.

Runs. Underwater paths, or trails, made by semiaquatic mammals as they perform their daily activities. Runs are most commonly seen between lodges or bank dens and feeding sites.

Rutting (the rut). The breeding season of deer, characterized in males by complete antler development and swollen necks.

S

Saprophages. Organisms that feed on dead organic matter.

Scats. The fecal droppings of animals.

Scrapes. Spots where leaf litter and soil have been pawed by a male deer advertising his presence during the rutting season.

Glossary

Selective cutting. Partial harvesting based on some selection criteria, e.g., timber harvesting or culling poor-quality trees.

Shade-intolerant. Referring to plants that require full sunlight and, therefore, are not capable of growing in full shade.

Shade-tolerant. Referring to plants that are capable of growing in full shade and, therefore, do not require full sunlight.

Siltation. The accumulation of silt in streams and ponds resulting from excessive soil erosion on banks and/or upland areas.

Singing grounds. Fields and clearings used by American woodcock for performing courtship flights, ground displays, and breeding.

Skid roads. Logging trails used for dragging (skidding) saw logs from the cutting area to the header (landing).

Slash. Branches and tops from trees that have been felled and bucked up, particularly after a logging operation.

Snags. Dead, standing trees. Snags are valuable to many wildlife species as sites for feeding and as sites in which to excavate or find cavities.

Snood. Fleshy growth protruding from the foreheads of turkeys.

Species. A group of individuals that look similar and are capable of breeding and producing viable offspring.

Substrate. Material that forms the bottom of wetlands, e.g., silt, sand, or cobble.

Subterminal band. The wide black or dark brown band near the tip of a ruffed grouse's tail.

Succulent vegetation. Herbaceous vegetation containing an abundance of water in its stems and leaves.

T

Territory. The area defended by an animal in attempt to "reserve" sufficient habitat resources to support itself, its mate and offspring, or a particular group of individuals.

Thinning. Selectively removing trees to improve stand quality for a specific objective. The trees to be removed are generally old (when thinning to maintain an uneven-aged stand at a desired age) or too dense or poor-quality (when thinning for timber improvements).

Timber- (sawtimber-) sized forests. Forest stands dominated by trees greater than 11 inches in diameter 4½ feet above ground.

Transit time. The time required for a particular nutrient to be recycled from a dead plant or animal back to the soil.

Trophic level. The location of an organism in the food pyramid or in a food chain.

U

Understory. Vegetation below the overstory in a forest.

Uneven-aged stands. Forest stands dominated by trees of different age classes.

W

Waterfowl. Ducks, geese, and swans.

Wetlands. As defined by Cowardin et al. 1979, "wetlands are lands transitional between terrestrial and aquatic systems where the water table is usually at or near the surface or [where] the land is covered by shallow water."

Wolf trees. Large, mature trees growing relatively free of competition from other trees. Wolf trees suppress the growth of timber trees, but they often provide food and cover for many animals.

Y

Yard. A densely vegetated, typically coniferous area where deer congregate in winter to find protection from the elements.

Literature Cited & Selected References

Numerous publications are available from local Cooperative Extension Centers, and many useful reference books can be obtained at libraries or book stores. The various field guide series are especially valuable for identifying the species discussed throughout this publication.

1 Basic Forest Wildlife Ecology

Anderson, S. H. 1985. *Managing Our Wildlife Resources.* Charles E. Merrill Publishing Company, Columbus, OH.

Arms, K., and P. S. Camp. 1982. *Biology.* Second edition. Saunders College Publishing Company, New York, NY.

Bailey, J. A. 1984. *Principles of Wildlife Management.* John Wiley and Sons, New York, NY.

Beattie, M., C. Thompson, and L. Levine. 1983. *Working with Your Woodland.* University Press of New England, Hanover, NH.

Brown, J. H., and A. C. Gibson. 1983. *Biogeography.* The C. V. Mosby Company, St. Louis, MO.

Dasmann, R. F. 1981. *Wildlife Biology.* John Wiley and Sons, New York, NY.

Decker, D. J., J. W. Kelley, T. W. Seamans, and R. R. Roth. 1983. *Wildlife and Timber from Private Lands: A Landowner's Guide to Planning.* Cornell Cooperative Extension. New York State College of Agriculture and Life Sciences, Cornell University, Ithaca, NY.

Minckler, L. S. 1975. *Woodland Ecology: Environmental Forestry for the Small Owner.* Syracuse University Press, Syracuse, NY.

Odum, E. P. 1983. *Basic Ecology.* Saunders College Publishing, Philadelphia, PA.

Richberger, W. E., and R. A. Howard, Jr. 1980. *Understanding Forest Ecosystems.* Cornell Cooperative Extension. New York State College of Agriculture and Life Sciences, Cornell University, Ithaca, NY.

Ricklefs, R. E. 1983. *The Economy of Nature.* Second edition. Chiron Press, Incorporated, Concord, MA.

Robinson, W. L., and E. G. Bolen. 1984. *Wildlife Ecology and Management.* Macmillian Publishing Company, New York, NY.

Smith, R. L. 1966. *Ecology and Field Biology.* Harper and Row, New York, NY.

2 Understanding Wildlife Habitats

Bailey, J. A. 1984. *Principles of Wildlife Management.* John Wiley and Sons, New York, NY.

Carey, A. B., and W. M. Healy. 1981. *Cavities in Trees around Spring Seeps in the Maple-Beech-Birch Forest Type.* Research Paper NE–480. USDI Forest Service. Northeastern Forest Experiment Station, Morgantown, WV.

Connecticut 208 Forestry Advisory Committee. 1982. *Logging and Water Quality in Connecticut: A Practical Guide for Protecting Water Quality while Harvesting Forest Products.* Connecticut Forest and Park Association, Incorporated, East Hartford, CT.

Cowardin, L. M., V. Carter, F. C. Golet, and E. T. LaRoe. 1979. *Classification of Wetlands and Deepwater Habitats of the United States.* USDI Fish and Wildlife Service, Biological Services Program FSW/OBS–79/31.

Dasmann, R. F. 1981. *Wildlife Biology.* John Wiley and Sons, New York, NY.

Decker, D. J., J. W. Kelley, and R. A. Howard, Jr. 1984. *Wildlife Habitat Enhancement.* Cornell Cooperative Extension. New York State College of Agriculture and Life Sciences, Cornell University, Ithaca, NY.

Decker, D. J., J. W. Kelley, T. W. Seamans, and R. R. Roth. 1983. *Wildlife and Timber from Private Lands: A Landowner's Guide to Planning.* Cornell Cooperative Extension. New York State College of Agriculture and Life Sciences, Cornell University, Ithaca, NY.

DeGraaf, R. M. 1984. *Managing New England Woodlands for Wildlife That Uses Tree Cavities.* Cooperative Extension Service, University of Massachusetts, Amherst, MA.

DeGraaf, R. M., and A. L. Shigo. 1985. *Managing Cavity Trees for Wildlife in the Northeast.* General Technical Report NE–101. United States Department of Agriculture, Forest Service. Northeastern Forest Experiment Station, University of Massachusetts, Amherst, MA.

Murie, O. J. 1974. *A Field Guide to Animal Tracks.* Houghton Mifflin Company, Boston, MA.

Robinson, W. L., and E. G. Bolen. 1984. *Wildlife Ecology and Management.* Macmillan Publishing Company, New York, NY.

Van Tyne, J., and A. J. Berger. 1976. *Fundamentals of Ornithology.* Second edition. John Wiley and Sons, New York, NY.

Whitaker, J. O. 1980. *The Audubon Society Field Guide to North American Mammals.* Alfred A. Knopf, New York, NY.

3 American Woodcock and Ruffed Grouse

Four comprehensive references regarding woodcock are as follows:

Liscinsky, S. A. 1972. *The Pennsylvania Woodcock Management Study.* Pennsylvania Game Commission Research Bulletin No. 171.

Owen, R. B., J. M. Anderson, J. W. Artmann, E. R. Clark, T. G. Dilworth, L. E. Gregg, F. W. Martin, J. D. Newsom, and S. R. Pursglove. 1977. "American Woodcock." Pages 149–186 in G. C. Sanderson, ed. *Management of Migratory Shore and Upland Game Birds in North America.* The International Association of Fish and Wildlife Agencies, Washington, D.C.

Sepik, G. F., R. B. Owen, and M. W. Coulter. 1981. *A Landowner's Guide to Woodcock Management in the Northeast.* USDI Fish and Wildlife Service Miscellaneous Report 253.

Sheldon, W. G. 1971. *The Book of the American Woodcock.* University of Massachusetts Press, Amherst, MA.

The following three references are comprehensive books pertaining to ruffed grouse:

Bump, B., R. W. Darrow, F. C. Edminster, and W. F. Crissey. 1947. *The Ruffed Grouse: Life History, Propagation, Management.* New York Conservation Department, Albany, NY.

Edminster, F. C. 1947. *The Ruffed Grouse.* The Macmillan Company, New York, NY.

Johnsgard, P. A. 1973. *Grouse and Quails of North America.* University of Nebraska Press, Lincoln, NE.

Many valuable publications are available from The Ruffed Grouse Society, 1400 Lee Drive, Coraopolis, PA 15108. Some of the more useful include:

Gullion, G. W. 1972. *Improving Your Forested Lands for Ruffed Grouse.*

Gullion, G. W. 1981. *The Ruffed Grouse.*

Gullion, G. W. 1983. *Managing Woodlots for Fuel and Wildlife.*

Gullion, G. W. 1984. *Managing Northern Forests for Wildlife.*

Gullion, G. W. Undated. *Improving Ruffed Grouse Habitat with Proper Planting.*

Most of the following technical references are obtainable at university and other large libraries or by contacting the authors. Only cited and selected references follow; no attempt has been made to list all of the numerous sources pertaining to American woodcock and ruffed grouse.

Almy, G. 1982. "Managing Forests for a Different Drummer." *American Forests* 88:16–21.

Berner, A., and L. W. Gysel. 1969. "Habitat Analysis and Management Considerations for Ruffed Grouse for a Multiple Use Area in Michigan." *Journal of Wildlife Management* 33:769–778.

Blankenship, L. H. 1957. *Investigations of the American Woodcock in Michigan.* Michigan Department of Conservation, Report No. 2123.

Brander, R. B. 1967. "Movements of Female Ruffed Grouse During the Mating Season." *Wilson Bulletin* 79:28–36.

Brown, C. P. 1946. "Food of Maine Ruffed Grouse by Seasons and Cover Types." *Journal of Wildlife Management* 10:17–28.

Chambers, R. E., and W. M. Sharp. 1958. "Movement and Dispersal within a Population of Ruffed Grouse." *Journal of Wildlife Management* 22:231–239.

Connors, J. I., and P. D. Doerr. 1982. "Woodcock Use of Agricultural Fields in Coastal North Carolina." Pages 139–147 in T. J. Dwyer and G. L. Storm, coords. *Woodcock Ecology and Management.* USDI Fish and Wildlife Service, Wildlife Research Report 14.

Gullion, G. W. 1967. "Selection and Use of Drumming Sites by Male Ruffed Grouse." *Auk* 84:87–112.

Gullion, G. W. 1977. "Forest Manipulation for Ruffed Grouse." *Transactions of the North American Wildlife and Natural Resources Conference* 42:449–458.

Gullion, G. W. 1985. "Aspen Management—An Opportunity for Maximum Integration of Wood Fiber and Wildlife Benefits." *Transactions of the North American Wildlife and Natural Resources Conference* 50:249–261.

Gullion, G. W., and A. A. Alm. 1983. "Forest Management and Ruffed Grouse Populations in a Minnesota Coniferous Forest." *Journal of Forestry* 81:529–532, 536.

Gullion, G. W., R. T. King, and W. H. Marshall. 1962. "Male Ruffed Grouse and Thirty Years of Forest Management on the Cloquet Forest Research Center, Minnesota." *Journal of Forestry* 60:617–622.

Gutzwiller, J. J., and J. S. Wakeley. 1982. "Differential Use of Woodcock Singing Grounds in Relation to Habitat Characteristics." Pages 51–54 in T. J. Dwyer and G. L. Storm, coords. *Woodcock Ecology and Management.* USDI Fish and Wildlife Service, Wildlife Research Report 14.

Kubisiak, J. F. 1978. *Brood Characteristics and Summer Habitats of Ruffed Grouse in Central Wisconsin.* Department of Natural Resources, Technical Bulletin No. 108. Madison, WI.

Lambert, R. L., and J. S. Barclay. 1976. "Woodcock Singing Grounds and Diurnal Habitat in North Central Oklahoma." *Proceedings: The Southeastern Association Game and Fish Commissioners* 29:617–630.

Larson, J. S., and R. D. Taber. 1980. "Criteria of Sex and Age." Pages 143–202 in S. D. Schemnitz, ed. *Wildlife Management Techniques Manual.* The Wildlife Society.

Lyons, T. E., and S. H. Broderick. 1986. *Managing Oak Forests for Fuelwood.* The University of Connecticut Cooperative Extension Service. College of Agriculture and Natural Resources, The University of Connecticut, Storrs, CT.

McDowell, R. D. 1975. "Fall Diets of Connecticut Ruffed Grouse." Pages 80–94 in D. DeCarli, ed. *Transactions of the Northeast Section,* The Wildlife Society.

Moulton, J. C. 1968. *Ruffed Grouse Habitat Requirements and Management Opportunities.* Department of Natural Resources, Research Report No. 36. Madison, WI.

Schemnitz, S. D. 1970. "Fall and Winter Feeding Activities and Behavior of Ruffed Grouse in Maine." *Transactions of the Northeast Wildlife Conference* 27:127–140.

Sharp, W. M. 1963. "The Effects of Habitat Manipulation and Forest Succession on Ruffed Grouse." *Journal of Wildlife Management* 27:664–671.

Smith, R. W., and J. S. Barclay. 1978. "Evidence of Westward Changes in the Range of the American Woodcock." *American Birds* 32(6):1122–1127.

Soil Conservation Service. 1973. *Woodcock: Planning Considerations and Specifications.* USDA, 645–21 CT-RI IV.

Svoboda, F. J., and G. W. Gullion. 1972. "Preferential Use of Aspen by Ruffed Grouse in Northern Minnesota." *Journal of Wildlife Management* 36:1166–1180.

Vermont Fish & Wildlife Department. 1986. *Model Habitat Management Guidelines for Deer, Bear, Hare, Grouse, Turkey, Woodcock, and Non-Game Wildlife.* Agency of Environmental Conservation, Montpelier, VT.

Wood, G. W., M. K. Causey, and R. M. Whiting, Jr. 1985. "Perspectives on American Woodcock in the Southern United States." *Transactions of the North American Wildlife and Natural Resources Conference* 50:573–585.

4 White-tailed Deer and Eastern Wild Turkey

Only cited and selected references follow; no attempt has been made to list all of the numerous sources pertaining to white-tailed deer and the eastern wild turkey.

White-tailed Deer

Behrend, D. F., and R. D. McDowell. 1967. "Antler Shedding among White-tailed Deer in Connecticut." *Journal of Wildlife Management* 31(3):588–590.

Caslick, J. W., and D. J. Decker. 1979. "Economic Feasibility of a Deer-proof Fence for Apple Orchards." *Wildlife Society Bulletin* 7(3):173–175.

Connecticut Wildlife Bureau. 1985. *The White-tailed Deer.* Department of Environmental Protection, Hartford, CT. Informational Series TA–S–10.

Dickinson, N. R. 1972. *Deer Management Considerations in Forest Management.* Vermont Fish & Game Department, Agency of Environmental Conservation, Montpelier, VT.

Lyons, T. E., and S. H. Broderick. 1986. *Managing Oak Forests for Fuelwood.* University of Connecticut Cooperative Extension Service. College of Agriculture and Natural Resources, University of Connecticut, Storrs, CT.

Marchinton, R. L., and D. H. Hirth. 1984. "Behavior." Pages 129–168 in L. K. Halls, ed. *White-tailed Deer: Ecology and Management.* Wildlife Management Institute. Stackpole Books, Harrisburg, PA.

Mattfeld, G. F. 1984. "Northeastern Hardwood and Spruce/Fir Forests." Pages 305–330 in L. K. Halls, ed. *White-tailed Deer: Ecology and Management.* Wildlife Management Institute. Stackpole Books, Harrisburg, PA.

McCaffery, R. K., and W. A. Creed. 1969. *Significance of Forest Openings to Deer in Northern Wisconsin.* Department of Natural Resources. Technical Bulletin 44. Madison, WI.

Rue, L. L., III. 1978. *The Deer of North America.* Outdoor Life, New York, NY.

Smith, R. L., and J. L. Coggin. 1984. "Basis and Role of Management." Pages 571–600 in L. K. Halls, ed. *White-tailed Deer: Ecology and Management.* Wildlife Management Institute. Stackpole Books, Harrisburg, PA.

Vermont Fish & Game Department. 1979. *A Landowner's Guide: Wildlife Habitat Management for Vermont Woodlands.* Agency of Environmental Conservation, Montpelier, VT.

Weber, S. J. 1986. "White-tailed Deer Model Habitat Management Guidelines in Vermont." Pages 29–41 in *Model Habitat Management Guidelines for Deer, Bear, Hare, Grouse, Turkey, Woodcock, and Non-game Wildlife.* Vermont Fish & Wildlife Department, Agency of Environmental Conservation, Montpelier, VT.

Youatt, W. G., L. J. Verme, and D. E. Ullrey. 1965. "Composition of Milk and Blood in Nursing White-tailed Does and Blood Composition of Their Fawns." *Journal of Wildlife Management* 29:79–84.

Wild Turkeys

Davis, H. E. 1949. *The American Wild Turkey.* Small-Arms Technical Publishing Company, Georgetown, SC.

Davis, J. R. 1981. "Movements of Wild Turkeys in Southwestern Alabama." Pages 135–139 in Sanderson, G. C., and H. C. Schultz, eds. *Wild Turkey Management: Current Problems and Programs.* University of Missouri Press, Columbia, MO.

Grenon, A. G. 1986. *Habitat Use by Wild Turkeys Reintroduced in Southeastern Michigan.* M.S. Thesis. School of Natural Resources, The University of Michigan, Ann Arbor, MI.

Healy, W. M. 1981. "Habitat Requirements of Wild Turkeys in the Southeastern Mountains." Pages 24–34 in Bromley, P. T., and R. L. Carlton, eds. *Proceedings of the Symposium: Habitat Requirements and Habitat Management for the Wild Turkey in the Southeast*. Virginia Wild Turkey Federation, Richmond, VA.

Healy, W. M. 1985. "Turkey Poult Feeding Activity, Invertebrate Abundance, and Vegetation Structure." *Journal of Wildlife Management* 49:466–472.

Hurst, G. A., and C. N. Owen. 1980. "Effects of Mowing on Arthropod Density and Biomass as Related to Wild Turkey Brood Habitat." Pages 225–232 in J. M. Sweeney, ed. *Proceedings: Fourth National Wild Turkey Symposium*. Arkansas Chapter, The Wildlife Society.

Latham, R. M. 1956. *Complete Book of the Wild Turkey*. The Stackpole Company, Harrisburg, PA.

Lewis, J. C. 1973. *The World of the Wild Turkey*. Lippincott, Philadelphia, PA.

Miller, B. K. 1985. *The Connecticut Wild Turkey Program*. CT Department of Environmental Protection, Wildlife Bureau, Hartford, CT.

Miller, J. E., and H. L. Holbrook. 1984. *Return of a Native: The Wild Turkey Flourishes Again*.

Mosby, H. S., and C. O. Handley. 1943. *The Wild Turkey in Virginia: Its Status, Life History, and Management*. Virginia Commission of Game and Inland Fisheries, Richmond, VA.

Sanderson, G. C., and H. J. C. Schultz, eds. 1973. *Wild Turkey Management: Current Problems and Programs*. Missouri Chapter of the Wildlife Society. University of Missouri Press, Columbia, MO.

Schorger, A. W. 1966. *The Wild Turkey: Its History and Domestication*. University of Oklahoma Press, Norman, OK.

Speake, D. W. 1980. "Predation on Wild Turkeys in Alabama." *Proceedings: The National Wild Turkey Symposium* 4:86–101.

Speake, D. W., R. Metzler, and J. McGlincy. 1985. "Mortality of Turkey Poults in Northern Alabama." *Journal of Wildlife Management* 49:472–474.

5 Other Upland Forest Wildlife Species

Ashbaugh, B. L. 1965. *Trail Planning and Layout*. National Audubon Society, New York, NY.

Capen, D. E. 1979. "Management of Northeastern Pine Forests for Nongame Birds." Pages 00–100 in R. M. DeGraaf and K. E. Evans, comps. *Workshop Proceedings: Management of North Central and Northeastern Forests for Nongame Birds*. USDA Forest Service General Technical Report NC–51.

Capen, D. E. 1982. "Ecological Aspects of Managing Forest Bird Communities." Pages 23–31 in R. J. Regan and D. E. Capen, eds. *Conference Proceedings: The Impact of Timber Management Practices on Nongame Birds in Vermont*. The Wildlife Habitat Improvement Program, Vermont Fish & Wildlife Department, Agency of Environmental Conservation, Montpelier, VT.

Carey, A. B., and W. M. Healy. 1981. *Cavities in Trees around Spring Seeps in the Maple-Beech-Birch Forest Type*. USDA Forest Service General Technical Report NE–480.

Chadwick, N. L, D. R. Progulske, and J. T. Finn. 1986. "Effects of Fuelwood Cutting on Birds in Southern New England." *Journal of Wildlife Management* 50(3):398–405.

Chesnut, M. 1987. "Prescribed Burning for More Wildlife." *Outdoor Oklahoma* 43(1):40–45.

Connecticut RC&D Forestry Committee. 1990. *Timber Harvesting and Water Quality in Connecticut: A Practical Guide for Protecting Water Quality While Harvesting Forest Products*.

Decker, D. J., J. W. Kelley, and R. A. Howard, Jr. 1984. *Wildlife Habitat Enhancement*. Cornell Cooperative Extension. New York State College of Agriculture and Life Sciences, Cornell University, Ithaca, NY.

Decker, D. J., J. W. Kelley, T. W. Seamans, and R. R. Roth. 1983. *Wildlife and Timber from Private Lands: A Landowner's Guide to Planning*. Cornell Cooperative Extension. New York State College of Agriculture and Life Sciences, Cornell University, Ithaca, NY.

DeGraaf, R. M. 1982. "Breeding Bird Assemblages in New England Northern Hardwoods." Pages 5–22 in R. J. Regan and D. E. Capen, eds. *Conference Proceedings: The Impact of Timber Management Practices on Nongame Birds in Vermont*. The Wildlife Habitat Improvement Program, Vermont Fish & Wildlife Department, Agency of Environmental Conservation, Montpelier, VT.

DeGraaf, R. M. 1984. *Managing New England Woodlands for Wildlife That Uses Tree Cavities*. Cooperative Extension Service. University of Massachusetts, Amherst, MA.

DeGraaf, R. M., and A. L. Shigo. 1985. *Managing Cavity Trees for Wildlife in the Northeast*. USDA Forest Service General Technical Report NE–101.

DeGraaf, R. M., and D. D. Rudis. 1986. *New England Wildlife: Habitat, Natural History, and Distribution*. USDA Forest Service General Technical Report NE–108.

Dudderer, G. 1987. "How Harvest Methods Affect Wildlife." *American Forests* 93(9 & 10):21–23, 77.

Etling, K. 1987. "Private Forests: Crop Trees 'n Critters." *American Forests* 93 (9 & 10):21–22, 24, 74–77.

Evans, K. E., and R. N. Conner. 1979. "Snag Management." Pages 214–225 in R. M. DeGraaf and K. E. Evans, comps. *Workshop Proceedings: Management of North Central and Northeastern Forests for Nongame Birds*. USDA Forest Service General Technical Report NC–51.

Gill, J. D. 1982. "Forest Management for Cavity-nesting Birds." Pages 40–47 in R. J. Regan and D. E. Capen, eds. *Conference Proceedings: The Impact of Timber Management Practices on Nongame Birds in Vermont*. The Wildlife Habitat Improvement Program, Vermont Fish & Wildlife Department, Agency of Environmental Conservation, Montpelier, VT.

Godin, A. J. 1977. *Wild Mammals of New England*. Johns Hopkins University Press, Baltimore, MD.

Hassinger, J., C. E. Schwarz, and R. G. Wingard. 1981. *Timber Sales and Wildlife*. Pennsylvania Game Commission, Pennsylvania Department of Environmental Resources, Pennsylvania State University, and USDA Forest Service.

Hassinger, J., L. Hoffman, M. J. Puglisi, T. D. Rader, and R. G. Wingard. 1979. *Woodlands and Wildlife: Making Your Property Attractive to Wildlife*. College of Agriculture, The Pennsylvania State University, University Park, PA.

Hynson, J., P. Adamus, S. Tibbetts, and R. Darnell. 1982. *Handbook for Protection of Fish and Wildlife from Construction of Farm and Forest Roads*. USDI Fish and Wildlife Service. FWS/OBS–82/18.

Probst, J. R. 1979. "Oak Forest Bird Communities." Pages 80–88 in R. M. DeGraaf and K. E. Evans, comps. *Workshop Proceedings: Management of North Central and Northeastern Forests for Nongame Birds*. USDA Forest Service General Technical Report NC–51.

Royar, K. J. 1986. "Snowshoe Hare Model Habitat Guidelines in Vermont." Pages 6–13 in *Model Habitat Management Guidelines for Deer, Bear, Hare, Grouse, Turkey, Woodcock, and Non-Game Wildlife*. Vermont Fish & Wildlife Department, Agency of Environmental Conservation, Montpelier, VT.

Shissler, B. P., and M. E. Holman. 1987. "Ten Pitfalls to Avoid." *American Forests* 93 (9 & 10):25, 65, 67.

Temple, S. A., M. J. Mossman, and B. Ambuel. 1979. "The Ecology and Management of Avian Communities in Mixed Hardwood-Coniferous Forests." Pages 132–153 in R. M. DeGraaf and K. E. Evans, comps. *Workshop Proceedings: Management of North Central and Northeastern Forests for Nongame Birds*. USDA Forest Service General Technical Report NC–51.

Thomas, J. W., tech. ed. 1979. *Wildlife Habitats in Managed Forests: The Blue Mountains of Oregon and Washington*. USDA Agriculture Handbook 553.

Vermont Agency of Environmental Conservation. Undated. *Guides for Controlling Soil Erosion and Water Pollution on Logging Jobs in Vermont*.

Weber, S. J. 1986. "Nongame Forest Habitat Management Guidelines in Vermont." Pages 42–51 in *Model Habitat Management Guidelines for Deer, Bear, Hare, Grouse, Turkey, Woodcock, and Non-game Wildlife*. Vermont Fish and Wildlife Department, Agency of Environmental Conservation, Montpelier, VT.

6 Wetlands Wildlife

Many of the following technical and nontechnical references are available from university and town libraries or by contacting the authors or agencies.

Ammann, A. P., R. W. Franzen, and J. L. Johnson. 1986. *Method for the Evaluation of Inland Wetlands in Connecticut*. Connecticut Department of Environmental Protection, Bulletin No. 9.

Bellrose, F. C. 1976. *Ducks, Geese, and Swans of North America*. Stackpole Books, Harrisburg, PA.

Bookhout, T. A., ed. 1977. *Symposium Proceedings: Waterfowl and Wetlands—An Integrated Review*. LaCrosse Printing Company, LaCrosse, WI.

Brinson, M. M., B. L. Swift, R. C. Plantico, and J. S. Barclay. 1981. *Riparian Ecosystems: Their Ecology and Status*. USDI Fish and Wildlife Service, Biological Service Program FWS/OBS–81/17.

Connecticut RC&D Forestry Committee. 1990. *Timber Harvesting and Water Quality in Connecticut: A Practical Guide for Protecting Water Quality While Harvesting Food Products*.

Cowardin, L. M., G. E. Cummings, and P. B. Reed, Jr. 1967. "Stump and Tree Nesting by Mallards and Black Ducks." *Journal of Wildlife Management* 31:229–235.

Cowardin, L. M., V. Carter, F. C. Golet, and E. T. LaRoe. 1979. *Classification of Wetlands and Deepwater Habitats of the United States*. USDI Fish and Wildlife Service, Biological Service Program FWS/OBS–79/31.

Craven, S., G. Jackson, W. Swenson, and B. Webendorfer. 1987. *The Benefits of Well-Managed Stream Corridors*. University of Wisconsin Extension, Madison, WI.

Damman, A. W. H., and T. W. French. 1987. *The Ecology of Peat Bogs of the Glaciated Northeastern United States: A Community Profile*. USDI Fish and Wildlife Service, Biological Report 85(7.16).

Deems, E. F., Jr., and D. Pursley, eds. 1978. *North American Furbearers: Their Management, Research, and Harvest Status in 1976*. International Association of Fish and Wildlife Agencies. University Press, College Park, MD.

DeGraaf, R. M., and D. D. Rudis. 1986. *New England Wildlife: Habitat, Natural History, and Distribution*. USDA Forest Service General Technical Report NE–108.

Errington, P. L. 1961. *Muskrats and Marsh Management*. The Wildlife Management Institute. Stackpole Books, Harrisburg, PA.

Godin, A. J. 1977. *Wild Mammals of New England*. The Johns Hopkins University Press, Baltimore, MD.

Harris, L. D. 1985. *Conservation Corridors: A Highway System for Wildlife*. Florida Conservation Foundation, ENFO.

Hassinger, J., C. E. Schwarz, and R. G. Wingard. 1981. *Timber Sales and Wildlife*. Pennsylvania Game Commission, Harrisburg, PA.

Heusmann, H. W. 1982. "Mallard-Black Duck Relationships in the Northeast." Pages 1249–1255 in Ratti, J. T., L. D. Flake, and W. A. Wentz, comps. *Waterfowl Ecology and Management: Selected Readings*. The Wildlife Society, Inc., Bethesda, MD.

Heusmann, H. W., and R. H. Bellville. 1982. *Wood Duck Research in Massachusetts 1970–1980*. Massachusetts Division of Fisheries and Wildlife. Research Bulletin 19.

Hynson, J., P. Adamus, S. Tibbetts, and R. Darnell. 1982. *Handbook for Protection of Fish and Wildlife from Construction of Farm and Forest Roads*. USDI Fish and Wildlife Service, Biological Service Program FWS/OBS–82/18.

Kundt, J. F., and T. Hall. 1988. *Streamside Forests: The Vital, Beneficial Resource*. University of Maryland Cooperative Extension, College Park, MD.

Laitin, J. 1987. "Corridors for Wildlife." *American Forests*. 93(9&10):47–49.

Linduska, J. P., and A. L. Nelson, eds. 1964. *Waterfowl Tomorrow*. USDI Fish and Wildlife Service, Government Printing Office, Washington, D.C.

MacClintock, L., R. F. Whitcomb, and B. L. Whitcomb. 1977. "Island Biogeography and 'Habitat Islands' of Eastern Forest. II. Evidence for the Value of Corridors and Minimization of Isolation in Preservation of Biotic Diversity." *American Birds* 31:6–16.

Mendall, H. L. 1958. *The Ring-necked Duck in the Northeast*. University of Maine Studies, Second Series, No. 73. University Press, Orono, ME.

Miller, S., and C. M. Pennock, and committees. 1979. *Best Management Practices Handbook: Forestry*. Virginia State Water Control Board, Planning Bulletin 317. Richmond, VA.

Newton, R. B. 1988. *Forested Wetlands of the Northeast*. Environmental Institute Publication No. 88–1, University of Massachusetts, Amherst, MA.

Niering, W. A. 1966. *The Life of the Marsh*. McGraw-Hill Book Company, New York, NY.

Niering, W. A. 1985. *Wetlands*. The Audubon Society. Alfred A. Knopf, Inc., New York, NY.

Pough, R. H. 1953. *Audubon Water Bird Guide: Water, Game, and Large Land Birds of Eastern and Central North America from Southern Texas to Central Greenland*. Doubleday and Company, Inc., Garden City, NY.

Peterson, R. T. 1980. *A Field Guide to the Birds: A Completely New Guide to All the Birds of Eastern and Central North America*. Fourth edition. Houghton Mifflin Company, Boston, MA.

Rude, K. 1987. "An Age-old Image Problem." *Ducks Unlimited* 51(5):80, 83, 158–159.

Shaw, S. P., and C. G. Fredine. 1956. *Wetlands of the United States*. USDI Fish and Wildlife Service Circ. 39.

Teskey, R. O., and T. M. Hinckley. 1977. *Impact of Water Level Changes on Woody Riparian and Wetland Communities. Vol. I: Plant and Soil Responses to Flooding*. USDI Fish and Wildlife Service, Biological Service Program FWS/OBS–77/58.

Teskey, R. O., and T. M. Hinckley. 1978. *Impact of Water Level Changes on Woody Riparian and Wetland Communities. Vol. II: Northern Forest Region*. USDI Fish and Wildlife Service, Biological Service Program FWS/OBS–78/88.

USDA. 1971. *Wild Ducks on Farmland in the South*. Farmer's Bulletin 2218.

Vermont Agency of Environmental Conservation. Undated. *Guides for Controlling Soil Erosion and Water Pollution on Logging Jobs in Vermont*.

Wade, D. A., and C. E. Ramsey. 1986. *Identifying and Managing Aquatic Rodents in Texas: Beaver, Nutria, and Muskrats*. Texas Agricultural Extension Service, The Texas A&M University.

Weller, M. W. 1978. "Management of Freshwater Marshes for Wildlife." Pages 267–284 in Good, R. E., D. F. Whigham, and R. L. Simpson, eds. *Freshwater Wetlands: Ecological Processes and Management Potential*. Academic Press, New York, NY.

Weller, M. W. 1987. *Freshwater Marshes: Ecology and Wildlife Management*. University of Minnesota Press, Minneapolis, MN.

Yoakum, J., W. P. Dasmann, H. R. Sanderson, C. M. Nixon, and H. S. Crawford. 1980. "Habitat Improvement Techniques." Pages 329–403 in Schemnitz, S. D., ed. *Wildlife Management Techniques Manual*. The Wildlife Society, Washington, D.C.